T0054951

THE CRADLE OF HUMANITY

Mark Maslin is Professor of Earth System Science at University College London. He is a Royal Society Wolfson Research Merit Scholar, a Royal Society Industry Fellow and a Fellow of the Royal Geographic Society. Maslin has published over 160 papers in journals such as *Science* and *Nature* on past and future climate change and its effects on the carbon cycle, human health, biodiversity, and human evolution. He is the author of ten books, including *Climate: A Very Short Introduction* (OUP, 2013), *Climate Change: A Very Short Introduction* (OUP, 2014), and the bestseller *The Human Planet* (Penguin, 2018), with Simon Lewis.

Praise for *The Cradle of Humanity*

'Understanding the emergence of our species from the unique landscapes of East Africa is one of the great scientific challenges. Mark Maslin takes us on an exhilarating intellectual journey, encompassing geology, astronomy, climate science, and evolutionary biology, to argue that the unique landscape and ever-changing climate of the East African Rift Valley were instrumental in catalysing the emergence of a civilisation on our planet. I'm left with a dizzying feeling of our good fortune to be here at all, and a powerful sense of our responsibility, as Maslin notes, to earn our species name: "Wise".'

Professor Brian Cox

'In this tale of mountains, monsoons and meteorites, climate, and ocean currents, Maslin masterfully puts human evolution into context, and shows how the earth and its environments have shaped us.'

Professor Alice Roberts, anthropologist,
author, and broadcaster

'As we confront rapid, major changes in the earth's climate today, it is imperative we understand how past climate change made us who we are. This fast-paced book vividly tells the story of how and why shifting environments have been driving human evolution ever since our earliest beginnings in Africa, and why those changes matter.'

Professor Daniel E Lieberman, Harvard University
Author, *Story of the Human Body: Evolution, Health and Disease*

'This book offers far more than a palaeoanthropological cocktail with a twist . . . In synthesising the most recent research in palaeoanthropology and giving the ecology of our ancestors a climatological twist, Maslin has produced a book that is fascinating, humbling and informative.'

Adrian Barnett, *New Scientist*

'A powerful, gripping account of how the dynamic earth shaped human evolution . . . With impressive ease, Maslin packs a tremendous amount of knowledge into a flowing narrative, making the point that special conditions for a number of species of tropical apes on the African continent eventually turned out to be luck . . . A tour de force through Earth's history and a timely reminder of just how lucky we are to be here at all.'

Peter C. Kjærgaard, Director and Professor,
Natural History Museum of Denmark

'Anyone who reads *The Cradle of Humanity* will certainly be enlightened about this awe-inspiring journey.'

Andrew Robinson, *Current World Archaeology*

'Impressively in-depth and well-explained mix of encyclopaedic information . . . There is an amazing amount of information packed into this surprisingly slim book.'

Chris Fitch, *Geographical*

'Palaeoclimatologist Mark Maslin delves into deep time to trace humanity's rise to geological hegemony. Examining early hominin finds in East Africa, he spotlights three stages (bipedalism in Australophithecus, a jump in brain size in Homo erectus and Homo sapiens' arrival some 195,000 years ago) and the roles of climate change, celestial mechanics and plate tectonics in their emergence. Ultimately, he theorizes that 'climate pulses' in the Rift Valley, in which hyper-arid conditions alternated with the formation of vast lakes, helped to drive the evolution of the big hominin brain.'

Nature, Jan 2017

THE **CRADLE**
OF **HUMANITY**

How the changing landscape
of Africa made us so smart

MARK MASLIN

OXFORD
UNIVERSITY PRESS

OXFORD

UNIVERSITY PRESS

Great Clarendon Street, Oxford, OX2 6DP,
United Kingdom

Oxford University Press is a department of the University of Oxford.
It furthers the University's objective of excellence in research, scholarship,
and education by publishing worldwide. Oxford is a registered trade mark of
Oxford University Press in the UK and in certain other countries

© Eco-Climate Limited 2017

The moral rights of the author have been asserted

First published in hardback 2017
First published in paperback 2019

Impression: 1

Published in the United States of America by Oxford University Press
198 Madison Avenue, New York, NY 10016, United States of America

British Library Cataloguing in Publication Data

Data available

Library of Congress Cataloging in Publication Data

Data available

ISBN 978–0–19–870452–2 (Hbk.)
ISBN 978–0–19–870453–9 (Pbk.)

Printed and bound in Great Britain by
Clays Ltd, Elcograf S.p.A.

Links to third party websites are provided by Oxford in good faith and
for information only. Oxford disclaims any responsibility for the materials
contained in any third party website referenced in this work.

Dedicated to Professor Martin Trauth, without whom none
of this work on human evolution would have been possible.

FOREWORD

The Cradle of Humanity is an important contribution to the extensive literature on the subject of human evolution. Sadly there is no absolute truth to be offered as to what and who became what and whom; any number of interpreters have published their views, and sometimes even their arguments and reasons for holding such views. The point that emerges is that data have yet to be found, some may never be found, but each year, the aggregate of knowledge is more and more persuasive. We humans are indeed the result of biological evolution and those facts exist in museums and research collections. I stress this because even now in this century, countless allegedly learned men and women, active in science, speak of the 'theory of human evolution'. Surely—and this book fully attests to this—the evidence for evolution such as the fossils, dates, archaeology, and countless re-evaluations leading to refinements in the data sets are now so comprehensive and consistent that the idea of the 'theory' can be discarded entirely and permanently.

The multidisciplinary fields of investigation are primarily concerned with the analysis of the *facts* of evolution. For example, how much can investigators learn regarding why, when, and how our evolution has happened? We know for example that we had small-brained early ancestors and later, larger brained ones. The relationship between body mass and brain size grew over several million years. This is known, but why and in which populations is yet to be fully understood. Today we are fully bipedal and we use our hands (in conjunction with our brains) to manipulate objects rather than using them to support our bodies. Upright posture puts huge strains on the human frame (back problems, hip joints, etc.) and yet it is clearly advantageous. What pressures in the environment led to the shift from four-legged to two-legged locomotion? Explanations abound, but the fact is it happened, and this is important. Similarly, the question of the role that speech played in our becoming us: when, why, and where? These are questions which are important. Was the process gradual or rapid?

Professor Mark Maslin does not dwell in depth on these and many other evolutionary milestones concerning us, but what he has done, and this makes the book so interesting, is to put our evolution into a larger picture—a context that is increasingly understood but seldom talked about in popular texts. For many, the relationship between our place in the Earth's history and the history of our planet itself makes the story so much more acceptable. The third, fourth, fifth, and sixth chapters deal with our Earth and its history. The chapters discuss temperature gradients, ocean currents, rainfall patterns, and much more on a global scale. Many will find the account explaining mountain building, lifting, plate tectonics, and their impact or influence on ecology, changes in vegetation, and sedimentation useful when thinking about why, over time, life forms have had to change. Most of this discussion is written to explain Africa's, and in particular East Africa's special circumstances for favouring, or one might say driving, human evolution.

I have found that many of the concerned questions that have been put to me over forty years of public lecturing are now answered in this one book. This is of great value to students and the broader swathe of public interest. That Maslin's account is in some chapters different from other books should not be a worry for the reader; it illustrates the complexities of science and the inevitable debates arising from a variety of ideas to explain the hard facts. There are countless fossil specimens now known that simply make doubting our evolution an exercise in self-delusion. Read this excellent narrative and see for yourself.

Richard Leakey, FRS
4 March 2016

ACKNOWLEDGEMENTS

I would like to thank Johanna, Alexandra, and Abbie Maslin for being there for me. I would also like to thank my colleagues, who have put up with me continually asking questions, and who helped me on my journey of discovery within the wonderfully contentious field of human evolution. In random order they are: Beth Christensen, Annett Junginger, Alice Roberts, Susanne Shultz, Katie Wilson, Kit Opie, Leslie Aeillo, Craig S. Feibel, Robin Dunbar, Brian Cox, Richard Leakey, Djuke Veldhuis, Rick Potts, Adam Rutherford, Peter C. Kjærgaard, Alan Deino, Matt Pope, Kaye Reed, Chris Stringer, Martin Trauth, Fred Grine, Meave Leakey, Rob Foley, Sam Nicholson, David Price, Phil Hopley, Fiona Scott, Victoria Ling, Marta Lahr, Peter de Menocal, Janet Hope, Helen Douglas Dufresne, Pete Ilsley, John Kingston, Frank Schäbitz, Simon Conway-Morris, Asfawossen Asrat, Chronis Tzedakis, Matt Sponheimer, Fiona Jones, Verena Foerster, Julia Lee-Thorp, Jim Rose, Chris Brierley, Patrice Désilets, Jean-François Boivin, Marianne Lupien, and the other two Marks 'Thomas' and 'Collard'. I would also like to thank Latha Menon and Jenny Nugee for the excellent editing, nagging, and support, all the staff and friends at UCL who make it the best place to work, and Peter Kjærgaard and Martin Trauth for reviewing the manuscript. And last, but not least, Miles Irving for his excellent and distinctive illustrations.

CONTENTS

LIST OF FIGURES

Period (15–5 ka BP). *Global and Planetary Change* 111: 174–88, with
permission from Elsevier and thanks to the authors.

Adapted from original unpublished files for figure 1 in
Junginger, A. and Trauth, M.H. 2013. Hydrological constraints of
paleo-Lake Suguta in the Northern Kenya Rift during the African
Humid Period (15–5 ka BP). *Global and Planetary Change* 111: 174–88,
with permission from Elsevier and thanks to the authors.

Based on figure 6 in Maslin, M.A., Brierley, C.M., Milner, A.M.,
Shultz, S., Trauth, M.H., and Wilson, K.E. 2014. East African climate
pulses and early human evolution. *Quaternary Science Reviews*, 101: 1–17,
CC-BY-SA-3.0 <https://creativecommons.org/licenses/by/3.0>.

Adapted from figure 2 in Veldhuis, D., Kjærgaard, P.C., and
Maslin, M. 2014. Human evolution: theories and progress, in
Smith, C. (ed.) *Encyclopedia of Global Archaeology*, Springer,
New York, pp. 3520–32. With permission of Springer.

© Ilmari Karonen/Wikimedia Commons/Permission is granted to
copy, distribute and/or modify this document under the terms of
the GNU Free Documentation License, Version 1.3 or any later version
published by the Free Software Foundation; with no Invariant
Sections, no Front-Cover Texts, and no Back-Cover Texts. A copy
of the license is included in the section entitled 'GNU Free
Documentation License' <https://en.wikipedia.org/wiki/Wikipedia:
Text_of_the_GNU_Free_Documentation_License>.

1

Introduction

Humans are rather weak when compared with many other animals. We are not particularly fast and have no natural weapons. But we have become the world's apex predator and have taken over the planet. We *Homo sapiens* currently number over 7.5 billion and are set to rise to nearly 10 billion by the middle of this century. We have influenced almost every part of the Earth system and, as a consequence, are changing the global environmental and evolutionary trajectory of the Earth. We are also both inadvertently and intentionally changing our own evolution. Fundamental to our success is that we are very smart, both individually, but more importantly, collectively. But why did evolution favour the brainy ape? Because, let's face it, there are lots of drawbacks to having a large brain. First, every child tries to kill their mother, as it is difficult and dangerous to give birth to offspring with such a large head. Second, you need a large amount of food in order to keep the organ running, as big brains are energetically expensive. Third, it frequently goes wrong. So the large brain must have given our ancestors a major advantage.

This book pieces together the evidence for human evolution by focusing on the major evolutionary changes that occurred in Africa that led our ancestors to walk upright and then to become progressively smarter. Without the unique combination of tectonics, global and local environmental changes, and celestial mechanics, human evolution would never have occurred. In this introductory chapter I briefly review the origin of our planet to put our evolution in the context of all evolution on Earth. I then set out the ten transitions that have occurred on the way to humans becoming the new geological superpower. What I hope this book will show is that this human ascendancy was not inevitable but caused but a unique combination of factors. As the great evolutionary theorist Stephen J. Gould suggested, if we re-ran the tape of evolution again and again the chances of humans evolving would be zero.

In the beginning

To comprehend human existence and how we got here I believe we need to put our evolution into context. To do this we need to understand the origins of our planet and how life has evolved. So let us start right at the beginning of everything. The Universe, according to cosmologists, is 13.8 billion years old. It all started with the Big Bang, when all the matter in the Universe was created and blasted outwards. About 100,000 years after the creation of matter, its expansion allowed it to cool, so that negatively charged electrons were trapped by positively charged protons forming hydrogen gas. At this point there were no galaxies, no stars, no planets, and no life—only an expanding cloud of gas. Over time, this irregular cloud started to form clumps of matter. Gravity pulled on the gases within these clumps to form galaxies. Gravity within each galaxy pulled matter together to form billions and billions of stars.

Deep within each star the immense pressure resulting from gravity forces the hydrogen atoms to collide. These collisions occasionally lead to the creation of helium through the fusion of hydrogen atoms and results in a huge release of energy. Stars thus burn hydrogen through nuclear fusion. As a star's fuel is converted from hydrogen to helium, the helium starts to collide and its fusion creates carbon and oxygen. The carbon then burns to create oxygen, neon, sodium, and magnesium. This nuclear fusion or 'cooking' continues through the elements up to and including iron. These lighter elements are the main ones found in rocky planets and are often called the planet builders. Elements heavier than iron can only be created by neutron capture that involves adding energy. This only occurs in the ejecta of massive stars, ten to twenty-five times the size of the Sun, when they finally explode. Large stars that have massive gravitational force have the most intense nuclear fires and are extremely bright but relatively short-lived. These stars produce all the elements and distribute them throughout galaxies through supernova explosions at the end of their cosmically short lives. So we are all made of stardust. Though these massive stars create the building blocks for planets and life they themselves are not suitable for forming a habitable solar system. Smaller stars, like our Sun, have a smaller gravitational pressure and thus cook their hydrogen at a much slower rate and provide a more stable long-term environment to allow solar systems to develop. Our own Sun has already existed for 4.6 billion years; in 5 billion years' time it will expand to form a Red Giant, and will continue expanding until it runs out of fuel in about 7.5 billion years' time.

Our Solar System formed 4.568 billion years ago from the gravitational collapse of a giant interstellar gas cloud. The vast majority of the system's mass is in the Sun, with most of the remaining mass contained in Jupiter. There are four smaller inner or terrestrial planets—Mercury, Venus, Earth, and Mars. The four outer planets are giant planets. Then there is the asteroid belt, made up of numerous irregularly shaped bodies and minor planets. The total mass of the asteroid belt is about 4 per cent that of the Moon, of which half is made up by the four largest asteroids. The two largest planets, Jupiter and Saturn, are gas giants, and are composed mainly of hydrogen and helium. It is the position and mass of Jupiter that has helped to create our relatively stable Solar System. It also influences the changing shape of the Earth's orbit around the Sun, and this has a profound effect on the Earth's climate. The two outermost planets, Uranus and Neptune, are ice giants, and are made of water, ammonia, and methane. Further out are rocky fragments, one of which is Pluto, which to my youngest daughter's dismay has been downgraded to a 'dwarf planet' or planetoid.

Central to the stability of the Earth is the Moon. The Moon formed about 4.53 billion years ago, some 30–50 million years after the origin of the Solar System. The prevailing hypothesis is that the Earth–Moon system formed as a result of a giant impact. A Mars-sized planet, which has been named Theia, hit the proto-Earth, increasing the mass of the Earth while blasting material into orbit around the proto-Earth, which accreted to form the Moon. The Moon is thus made of the lighter elements found in the surface of the proto-Earth. So the Moon does not have a metal core and its density is lighter than Earth's. The Moon acts as the Earth's gyroscope, so its axis of rotation in relation to the Sun has remained very similar for the last 4.5 billion years. As we will see later, the small wobbles of the axis of the rotation are enough to push the Earth into and out of the 'great ice ages'. About 100 million years after the formation of the Solar System the Earth degassed, creating an early atmosphere of nitrogen and carbon dioxide as well as significant amounts of water. The Earth overall has very low amounts of the essential volatiles such as water, carbon dioxide, and methane, which are essential for life. Only one molecule in 3 million on Earth is water, but through these early processes they have been concentrated at the surface. At about 4 billion years ago we have the first evidence of rock formation on Earth (Figure 1).

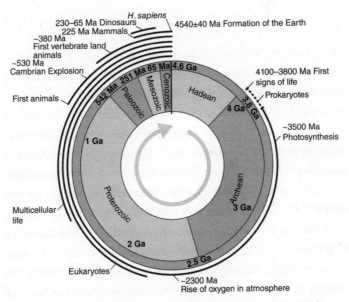

FIGURE 1 Major biological events in the 4.6 billion year history of the Earth.

Then, at about 3.5 billion years, we have the first evidence of life on Earth—and that is where our story really begins.

Ten key steps in human evolution

The essential requirements to form a habitable planet seem to be a small, long-lived star, a stable Solar System, and a planet at the right distance from the star so it is not too hot and not too cold, and which also has enough water and carbon at the surface to support life. Once these elements were established then there are ten key transitions that occurred, which inadvertently ended up with humans populating and dominating the whole planet. These are summarized below, and we will examine a number of them in more detail later.

1. *Origin of life*

Fossil evidence for life can be found as far back as 3.46 billion years ago in the form of preserved cellular structures in the Strelley Pool Formation in Western Australia. These are thought to be created by 'single-celled' bacterial communities that formed stromatolites and are still found today in the warm shallow seas in Shark Bay, Australia. Little seems to have happened during the following so-called 'boring' billion years until about 2.4 billion years ago, when a significant rise in atmospheric oxygen known as the Great Oxidation Event is recorded in the rocks. Until then, only simple, 'prokaryotic' organisms existed on Earth. The more complex 'eukaryotic' cells, with nuclei and other internal structures, emerged between 2.1 and 1.6 billion years ago from the merger of two types of prokaryotes, and these are the basis of almost all multicellular organisms.

2. *Origin of complex life*

The origin of animals is currently uncertain, but simple animals had probably evolved by about 600 million years ago. However, most of what were to become the major animal phyla we know today appear to have arisen during a period of rapid diversification known as the Cambrian Explosion, which began about 541 million years ago, the start of what is known as the Phanerozoic Eon, and lasted for over 40 million years. Some of the extraordinary forms that existed at this time are captured in the remarkably preserved fossil assemblages of the Burgess Shale (Rocky Mountains, Canada) and Chengjiang (Yunnan Province, China). This melting pot of evolution included one branch—the vertebrates—that evolved a backbone, and it is from these that we are descended.

3. *Evolution of mammals*

The unique features of mammals are milk-producing mammary glands in females, fur or hair, three bones in the inner ear which evolved from the reptilian jaw, and a neocortex (Figure 2). The neocortex is a region of the brain that controls higher functions such as sensory perception, generation of motor commands, spatial reasoning, and, in humans, conscious thought and language. The earliest mammals are thought to have emerged from mammal-like reptiles before the end of the Triassic Period, by about 225 million years ago.

4. *Extinction of the dinosaurs*

There have been several mass extinctions in Earth's history, the most extreme of which was at the end of the Permian Period, some 250 million years ago, which destroyed most of life on land and

Vertebrates
(525 million years ago)

Backbone
• a backbone

Mammals
(225 million years ago)

Fur
• fur or hair

Milk
• milk-producing glands

Ear
• three separate bones
  in the middle ear

Primates
(65 million years ago)

Nails
• fingernails and toenails

Thumb
• an opposable thumb...
  or big toe

Teeth
• four incisors in the upper
  and lower jaw

Hominids
(20–30 million years ago)
(apes/hominins)

Tail
• no tail

Shoulders
• shoulder blades at the back
  not at the sides

Teeth
• a Y-shaped pattern on the
  surface of the molars
  (chewing teeth)

FIGURE 2 Key morphological features that define hominids.

in the oceans. This has been referred to as the 'mother of all extinctions', but somehow life managed to survive and then diversify after this close call. The mass extinction most directly relevant to our story, however, is the Cretaceous–Tertiary extinction, also called the 'KT' extinction for short. Around 66 million years ago, a major period of volcanism producing massive outpourings of lava, resulting in the Deccan Traps in India, compounded by the Chicxulub meteorite impact event in Central America, caused a mass extinction that killed off many species, including the non-avian dinosaurs. This extinction event was essential in the story of human evolution as it ended the 170-million-year domination of dinosaurs and allowed the evolution and proliferation of mammals, and the appearance of the first ancestors of primates.

5. Evolution of social primates

Ten million years after the extinction of the dinosaurs, during a period of super-global warming, the first fossil evidence for true primates and social monkeys begins to appear. Anthropoids or simians, a group that includes primates and monkeys (Table 1), began to live in

TABLE 1. Common terminology used in human evolution and in this book

Anthropoids	All primates (monkeys and great apes and their fossil ancestors), hominins, and humans
Hominids	All great apes (gorillas, orangutans, chimpanzees, bonobos, gibbons), hominins, and humans
Hominins	All our fossil ancestors (*Ardipithecus, Australopithecus, Homo*)
Anatomically modern humans	*Homo sapiens*, but without substantial evidence for our cultural accoutrements (art, burials, ornament, musical instruments)
Humans	Only modern humans, *Homo sapiens*, with clear evidence of cumulative culture

larger groups, which meant that each animal had to negotiate complex webs of friendships, hierarchies, and rivalries. This is thought to have been a major driver of brain expansion later, in hominins.

6. Evolution of hominins

The exact origins of hominins is disputed. However, the last common ancestor with chimpanzees is thought to have lived between 8 and 5 million years ago. Despite the relatively miniscule genetic difference between modern humans and chimpanzees, our evolutionary paths have been radically different since the split.

7. Bipedalism

Between 10 and 5 million years ago, hominins developed the ability to move efficiently on two legs. At the same time chimpanzees and gorillas became better at tree-climbing and developed knuckle-walking while on the ground. Bipedalism allowed our ancestors to spread out from East Africa into Northern and Southern Africa. Some of these bipedal hominins were already using stone tools by at least 3.3 million years ago.

8. Brain expansion

About 2 million years ago new hominin species appeared with brains up to 80 per cent larger than their ancestors, and for the first time they dispersed out of Africa. This large brain was accompanied by other sweeping changes to life history (shortened intervals between births, delayed child development), body size, the shape of the pelvis, and a shoulder morphology which allowed projectile use. These species show adaptations to long-distance running, ecological flexibility, and social behaviour, including food processing.

9. Cumulative culture

Homo sapiens emerged about 200,000 years ago in East Africa and dispersed into Eurasia. It is not until 100,000 years ago that there is evidence of creative thinking. From 50,000 years ago this becomes more consistent, with evidence of art, ornaments, and symbolic behaviour. These steadily increase in complexity and frequency and demonstrate that knowledge was being generated and passed on to the next generation. Culture was being accumulated and grew with every generation.

10. Agriculture and industrial revolutions

At the end of the last ice age, about 11,000 years ago, agriculture first appeared in South West Asia, South America, and northern China. It then appeared 7,000–6,000 years ago in southern China and Central America and 5,000–4,000 years ago in the savannah regions of Africa, India, South East Asia, and North America. Agriculture steadily replaced hunter-gathering as the major economic mainstay, and urban centres and city-states developed. Humanity then went through the industrial, technological, and the information revolutions. Along the way the impact of humanity became so great that we emerged as a geological superpower and entered our own geological period: the 'Anthropocene'.

The rest of this book is focused on understanding why many of the later key transitions seem to occur in East Africa. We will consider how the environment in which our ancestors lived changed through time and how that may have affected their evolution. Plate tectonics, the movement of the Earth within the Solar System, and changes in global climate all appear to have played a part in our early evolution. Central to this whole story is the formation and development of the East African Rift Valley system, one of the most extensive geological features on the Earth's surface. It runs north–south for around 4,500 km from Syria through East Africa down to Mozambique. In Chapter 2, we examine the fossil evidence of our evolution, much of which has been found within the East African Rift Valley. This landscape is tough and unforgiving—but it appears to have been the cradle of humanity.

2

Early Human Evolution

To understand our evolutionary history it is essential that we understand the fossil record. Discoveries in the past few decades have dramatically expanded the available hominin fossil record, with thirteen new species suggested and four new genera named since 1987. This richer variety of fossils has produced two major changes in our knowledge. First, it has led to a much greater understanding of the range of variation in the hominin 'phenotype'—i.e. the observable characteristics or traits. This includes insights into populations, with fossils of multiple individuals found at Atapuerca (northern Spain), Dmanisi (Georgia), Hadar (Ethiopia), and Rising Star (South Africa). Second, the extensive use of new dating techniques has provided chronological precision to link those phenotypes to the environments in which they evolved. Even with all the recent finds the hominin fossil record is still very limited, with many gaps. There is also considerable discussion among researchers about defining these new species and genera, which influences our understanding of changes in overall hominin diversity.

But conflating or expanding the defined species seems at present to have little overall influence on the diversity pattern.

The other key debate concerns where all the new hominin species evolved. The date of first appearance of each species depends heavily on what happened after death, on the conditions in which the fossil was preserved, and on sampling bias. Nevertheless, the consistency of hominin first-appearance dates in East Africa currently suggests that the majority of the new hominin species evolved in East Africa and then dispersed outwards. This is supported by the present evidence for brain capacity, which suggests that brain expansion occurred first in East Africa and only appeared later elsewhere, following a dispersal event (see Figure 3). Some researchers suggest South African, European, or Asian origins for hominin speciation, but although these are valid possibilities there is currently no fossil evidence to support them. Genetic evidence also highlights the fact that there was a large amount of interbreeding between hominin species. This has led to a growing realization that the classical view of the branching hominin evolutionary tree is incorrect. Instead we should consider human evolution as more of a bush with a network of closely related species that frequently interbred, blurring the lines between species.

Five stages of human evolution

The fossil record suggests five main stages in hominin evolution: (1) the appearance of the earliest (proto) hominins attributed to the genera *Sahelanthropus, Orrorin*, and *Ardipithecus* between 7 and 4 million years ago; (2) the appearance of the *Australopithecus* genus around 4 million years ago; (3) the appearance of the *Homo* genus and the *Paranthropus* genus around the Plio–Pleistocene boundary

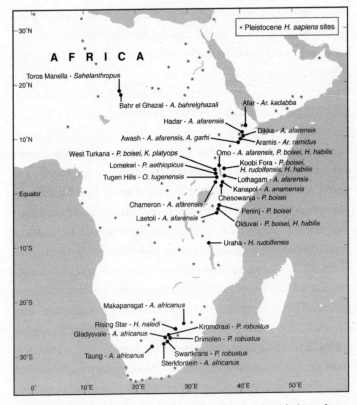

FIGURE 3 Map of locations of the most important hominin fossil finds in Africa.

between 2.8–2.5 million years ago; (4) the appearance of *Homo erectus* 1.8 million years ago; (5) the appearance of *H. heidelbergensis* 800,000 years ago, *Homo sapiens* 200,000 years ago, and modern humans around 50,000 years ago. The taxonomic classification of many specimens, as well as their role in the evolution of modern humans, is continually discussed, which is a polite way of saying continually fought over and argued about, both in public and in the literature. What is not disputed is that, apart from *Sahelanthropus* remains from Chad, all the earliest specimens for each of the main genera have been found in the East African Rift System.

Stage 1: Earliest hominins

The earliest potential candidate for the first hominin is *Sahelanthropus tchadensis*—though this is disputed. The fossils were discovered in the Djurab Desert of Chad by a team of four led by a Frenchman, Alain Beauvilain, and three Chadians, Adoum Mahamat, Djim-doumalbaye Ahounta, and Gongdibé Fanoné, and are dated to approximately 7–6 million years ago. When the discovery was announced in 2002, palaeoanthropologists were amazed by both the location and date. The skull fragments were found in a desert in Central Africa, far to the west of the other early hominin sites (Figure 3). The remains are limited to cranial fragments that suggest a mosaic of hominin and non-hominin features and a brain size equivalent to modern chimpanzees. What is striking about the skull, nicknamed 'Toumai' or 'hope of life', is how modern-looking it is, but with a thick brow ridge. The lack of other bones from the skeleton makes it extremely difficult to reconstruct its lifestyle, whether it was bipedal, or whether it was truly a hominin. However, the possible position and orientation of the '*foramen magnum*'—the hole in the base of the skull where the spinal cord passes through—strongly suggests a posture similar to other early bipeds (see Figure 4). The fossil was recovered from what is now a very dry desert area in Chad. The few teeth recovered suggest *Sahelanthropus* ate mainly fruit, which means Chad has undergone a large change in climate since *Sahelanthropus* was living there.

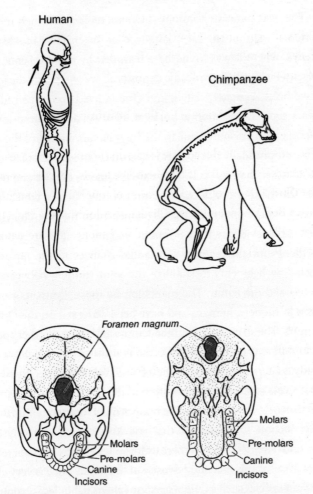

FIGURE 4 Location of the hole in the base of the skull through which the spinal cord passes (*foramen magnum*) in humans and chimpanzees, showing the central position essential for bipedal movement.

The next putative hominin is *Orrorin tugenensis*, which means 'original man' in the local language of Tugen Hills in western Kenya, where it was found by a team led by Brigitte Senut and Martin Pickford. The volcanic deposits above and below the fossil have been accurately dated, and *Orrorin* lived about 5.8 million years ago. The taxonomic position, lifestyle, and locomotion of *Orrorin* are all disputed due to the fragmentary nature of the specimens. Unusually this species is known for other parts of its skeleton rather than its skull. There are fragments of jaws and teeth, but the main evidence comes from two thighbones, which have preserved the upper end that is connected to the hip. This is the first direct evidence for a hominin walking upright on two legs. Walking upright requires substantial changes to the hip joint, including ligaments to stabilize the joint running between the pelvis and the femur. The marks left by these ligaments can be seen in modern humans and have been observed on the *Orrorin* femurs. The structure of human femurs and those of other apes is also different because when humans walk all the weight of the body is alternatively held by the left and right hip joints. Nearly all that stress is placed on the bottom of the femur neck, the narrow section below the ball—so the bone is much denser in humans at this section than in apes that use knuckle-walking to move around (Figure 5). This is where the disputes about *Orrorin* bipedalism arise, as this increased density of the femur bone is very clear in later species such as *Australopithecus afarensis*, but less obvious in the *Orrorin* femurs. *Orrorin* teeth are much more ape-like, though the teeth have thick enamel like humans. Hence, both *Sahelanthropus* and *Orrorin* have been assigned to hominins by their discoverers based on very different evidence—they may or may not be hominins, and even if they were hominins there is no

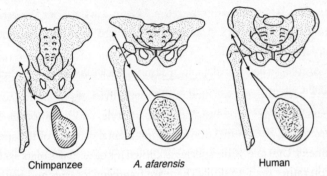

Chimpanzee   *A. afarensis*   Human

FIGURE 5 Evidence of bipedalism from the shape of the neck of the femur (thighbone).

way to tell if they were part of our ancestry. Both *Sahelanthropus* and *Orrorin* have been suggested as members of a clade that includes *Ardipithecus*—in other words, all three may have had a common ancestor which we are yet to find.

The oldest member of the *Ardipithecus* genus is *Ardipithecus kadabba*, whose fossil evidence consists only of fragmentary teeth and skeletal remains dated to approximately 5.6 million years ago. A much more extensive fossil record exists for the second member of the genus, *Ardipithecus ramidus*. The remains are from a female, nicknamed 'Ardi', and include most of the skull and teeth, as well as the pelvis, hands, and feet. Ardi was discovered by the palaeoanthropologist Yohannes Haile-Selassie, who was part of a team led by Tim White, in Ethiopia's harsh Afar desert, at a site called Aramis in the Middle Awash region. The name *Ardipithecus ramidus* stems mostly from the Afar language in which *Ardi* means 'ground or floor' (borrowed from either Amharic or Arabic) and *ramid* means 'root'. The *-pithecus* portion of the name is from the Greek word for 'ape'. The bones and teeth of nine other *A. ramidus* individuals have been discovered at As Duma in the Gona Western Margin of Ethiopia's Afar region, dated to between 4.35 and 4.45 million years old.

Ardipithecus had a small brain, measuring between 300 and 350 cm³. This is slightly smaller than a modern bonobo or a female common chimpanzee brain, and much smaller than the brain of *Australopithecines* like Lucy (~400 to 550 cm³). It is roughly 20 per cent the size of the modern human brain. *Ardipithecus* teeth indicate a highly omnivorous diet and their post-crania suggest a lifestyle of arboreality (tree dwelling) coupled with primitive bipedality. The fauna and vegetation associated with the *A. ramidus* specimens in the Awash Valley date to around 4.4 million years ago, suggesting a woodland-forest matrix habitat associated with significant rainfall and water availability. This appearance of bipedality in closed woodland environments undermines theories of bipedality evolving exclusively as an adaptation to open habitats. As we shall see, a number of researchers think that bipedalism may have evolved many times in the past because it proved useful for survival.

Stage 2: Appearance of *Australopithecus*

Australopithecus was first discovered in the 1920s and changed our understanding of human evolution. Because the *Australopithecus* species were small-brained, bipedal ancestors, this showed us that the defining feature of hominins was walking upright, not larger brains. *Australopithecus* means 'southern ape' and was coined by Raymond Dart when he discovered *A. africanus* in South Africa in 1924. When other similar fossils were found in East Africa the name was kept due to the close taxonomic similarity.

The first members of the *Australopithecus* genus, attributed to *A. anamensis*, were found at two sites near Lake Turkana, Kanapo and Allia Bay, and range in age from 4.2 to 3.8 million years ago. These individuals show strong evidence of bipedality combined with primitive cranial features. The species name *anamensis* comes from the Turkana word 'anam', meaning 'lake'. The sample of *A. anamensis* fossils is small, but a well-preserved tibia (shin) bone is very similar to a human tibia, suggesting bipedal locomotion.

A. anamensis is followed by *A. afarensis*, which is very well known from the fossil record and includes the remarkably complete 'Lucy' specimen. Lucy was found by a team led by Donald Johanson and Tom Grey on 24 November 1974 at Hadar in Ethiopia. The story goes that they had taken a Land Rover out that day to map another locality. After an unsuccessful morning surveying for fossils, they decided to head back to the vehicle. Johanson suggested taking an alternate route back to the Land Rover, through a nearby gully. Within moments he spotted a right proximal ulna (forearm bone) and identified it as a hominin bone. Over the next two weeks they recovered several hundred fragments of bone, representing 40 per cent of a single hominin skeleton—'Lucy'. How 'Lucy' got her name has gone down in folklore. On the night of 24 November there was much celebration and excitement over the discovery, with drinking, dancing, and singing, mainly to the Beatles' song Lucy in the Sky With Diamonds, which was played over and over again. At some point during that night, and no one has owned up to it, someone named the skeleton Lucy, and it has stuck. Our understanding of *A. afarensis* now goes way beyond Lucy, with another two complete skulls and a large collection of jaws and teeth, as well as a comprehensive collection of limb bones.

The species name *afarensis* means 'of the Afar region', where the fossil was initially found, and it has been dated to between 3.6 and 3 million years ago. *A. afarensis* still retains a small brain size, yet the post-cranial morphology is more similar to modern humans than to apes and suggests a lifestyle strongly adapted to long-distance walking. It seems to have been a highly successful species, and specimens have been found in Chad, Ethiopia, Kenya, and Tanzania. The jawbone found in Chad has been assigned the species name of *Australopithecus bahrelghazali*, partly to recognize a unique location and some different characteristics, but mainly as researchers desperately want to find new species and to name them. The *Australopithecus* genus may be wider still because in 2001 on the western side of Lake Turkana, Meave Leakey found what they named *Kenyanthropus platyops*. This skull was dated to 3.5 million years ago, and though similar to *A. afarensis* it has a very flat face—*platyops* means 'flat face', and is derived from two Greek words: *platus*, which means 'flat', and *opsis*, which means 'face'. Usually a flat face is associated with strong muscles for chewing, but there is no other evidence for this—for example, the teeth are small. Much of the rest of the skull's features are primitive and very similar to *A. afarensis*, so other researchers now refer to the species as *Australopithecus platyops*. Recent reanalysis of all the data has, however, convinced Meave Leakey and her colleagues that this is a separate species, and I will use *Kenyanthropus platyops* in this book.

The fossil evidence clearly indicates that *A. afarensis* was fully bipedal, and not quadrupedal like most monkeys and chimpanzees, nor suspensory like orangutans or gibbons. It did not run along the ground on all fours, or have long arms for climbing and swinging. This terrestrial bipedalism is shown by the beautiful and haunting fossilized footprint trail found in Laetoli in Tanzania, dated to 3.5 million years ago. But it is also clear that A. *afarensis* walked in a slightly different way to modern humans, as the blades on their hip bones flare out further to the sides. In summary, A. *afarensis* was a similar size to a modern chimpanzee, and was bipedal on the ground, but its relatively long arms and feet and the curved bones of its feet and fingers would have been used to climb trees for food, safety, and sleeping.

Recent finds in the Afar region of Ethiopia suggest that A. *afarensis* may in fact be two species. In 2011, three fossilized jawbones were found 35 km north of Hadar, where Lucy was found, in sediment dated to between 3.5 and 3.3 million years ago. Though the new fossils are similar to the Lucy A. *afarensis* specimen, the teeth have different root structures and are smaller. The cheekbones seem to be more forward-facing, the lower jaw is larger, and the teeth have thicker outer enamel than A. *afarensis*, suggesting adaptation to harder, tougher, and more abrasive food than A. *afarensis*. Yohannes Haile-Selassie, the curator of physical anthropology at the Cleveland Museum of Natural History, suggests these fossils may represent a new species—*Australopithecus deyiremeda*. The name 'deyiremeda' means 'close relative' in the local Afar language. If these fossils represent a new species, then some fossils identified as A. *afarensis* may belong to A. *deyiremeda*. But the naming and identification of A. *deyiremeda* as a new species is still controversial and requires a lot more evidence.

A. *afarensis* may hold another huge surprise for us. In 2011 and 2012 the West Turkana Archaeological Project found the oldest known stone tools, dated at 3.3 million years, on the western shore of Lake Turkana in northern Kenya. These tools predate by a million years the oldest fossil specimens attributed to *Homo* in West

Turkana at 2.35 million years ago, and by half a million years the earliest *Homo* find from Ledi-Geraru, Afar, Ethiopia, at 2.8 million years ago. This is revolutionary, as it has always been assumed that the evolution of *Homo* was associated with the start of stone tool-making. The only hominin species known to have been living in the West Turkana region at a similar time is *K. platyops* at about 3.5 million years ago, while in the Lower Awash Valley in Ethiopia, *A. afarensis* is found at 3.39 million years ago. The significance of these fossils is that they were found in association with cut-marked bones, indicating butchering. Bone from both impala and buffalo-sized creatures were found to bear cut marks. Both the stone tools and the cut-marked bones suggest that *A. afarensis* or *K. platyops* were not only making and using tools, but also venturing out of the safety of the forests and on to the plains in search of meat.

The oldest hominin found in South Africa is *Australopithecus africanus*, which simply means 'southern ape of Africa', (Figure 3) with the first appearance dated to 3.3 million years ago. It is similar to *A. afarensis* but with more ape-like limb proportions yet less primitive teeth. The longer femur in *A. afarensis*, as compared to *A. africanus*, suggests a longer stride and more efficient walking style. There are now hundreds of fossils associated with *A. africanus*, largely from the cave sites of Sterkfontein and Makapansgat in South Africa. Sterkfontein is close to Johannesburg, and is part of a set of fossil sites named the Cradle of Humankind World Heritage Site. *A. africanus* seems to have had a varied diet and the proportions of carbon isotopes in their teeth has shown that their diet may have contained either grass, grass seeds, or animals whose primary food was grass. *A. africanus* teeth and jaws seem to have no adaptations to eating grass or grass seeds. Given what we have recently found out about A. *afarensis* and stone tool use, we can speculate that A. *africanus* was also using stone tools to eat a significant amount of meat. Though we know that chimpanzees hunt meat, it makes up less than 5 per cent of their diet, but the carbon isotope evidence suggests A. *africanus* was eating a lot more than 5 per cent meat. A. *africanus* teeth are not adapted to slice meat effectively, so using stone tools makes perfect sense.

The final australopithecine is A. *garhi*, associated with 2.5 million-year-old deposits in the Awash Valley, Ethiopia (Figure 3). This is a confusing set of fossils as it consists of a few skull fragments and very large teeth; then, at a nearby site, a partial skeleton has been found with no skull fragments; and, at another nearby site, there are animal bones with evidence of cut marks. If these fossils do represent one individual then it was a long-legged creature with long forearms that may have cut the meat off bones of animals.

Stage 3: *Homo* and *Paranthropus*

In 1964, Louis Leakey and his colleagues announced the discovery of *Homo habilis* based on fossils found in the Olduvai Gorge in Tanzania (Figure 3). They described it as bipedal, a tool-maker, and with a larger brain than *Australopithecus*. The name *Homo habilis* means 'skilful man' (sometimes incorrectly translated as 'handy man'). *H. habilis* has been found throughout Eastern and Southern Africa. The age range for *H. habilis* is from 2.35 to 1.5 million years ago. In 2013, however, a fragment of a fossilized jawbone was found in the Ledi-Geraru research area, Afar, Ethiopia, by an Ethiopian student, Chalachew Seyoum. The fossil has been dated to 2.8 million years ago and is considered the earliest evidence of the *Homo* genus known to date. The new fossil seems to be intermediate between

Australopithecus and *H. habilis*. This discovery pushes back the origin of *Homo* by over 400,000 years.

H. habilis is a controversial species with much scholarly debate regarding its placement in the genus *Homo* rather than the genus *Australopithecus*. The small size and rather primitive attributes have led some experts, such as palaeoanthropologists Bernard Wood and Mark Collard, to propose excluding *H. habilis* from the genus *Homo* and placing it instead in *Australopithecus* as *Australopithecus habilis*. This is because, in its appearance and morphology, *H. habilis* is the least similar to modern humans of all species in the genus *Homo* (except the equally controversial *H. rudolfensis*). *H. habilis* was short and had disproportionately long arms compared to modern humans; however, it had a less protruding face than the *Australopithecines* from which it may have descended. *H. habilis* had a cranial capacity of between 550 and 687 cm^3, though a recent study led by anatomist and palaeoanthropologist Fred Spoor at University College London suggests one key specimen had an endocranial volume of between 729 and 824 cm^3, larger than any previously published value. So, compared to *Australopithecines*, *H. habilis'* brain capacity was on average 50 per cent larger, but it was still considerably smaller than the 1,350 to 1,450 cm^3 range of modern *Homo sapiens*. But what had until recently made *H. habilis* special was that, despite the ape-like morphology of their bodies, their remains were often accompanied by primitive stone tools at locations such as Olduvai Gorge in Tanzania and Lake Turkana in Kenya. The discovery of stone tools prior to the first appearance of *Homo* raises the question again of whether *Homo habilis* should be considered a member of the genus *Homo*.

The definition of *H. habilis* suffers from another problem: as more and more specimens were classified as *H. habilis*, there seemed to be too much physical variation to occur in one species. For example, many of the small specimens had more prominent brow ridges—usually found in larger males in modern species of apes. To get round this problem a new controversial species, *Homo rudolfensis*, was proposed to encompass the larger individual specimens with bigger brains and a slightly different facial structure and chewing mechanism. The species is named after Lake Rudolf, the former name for Lake Turkana, in Kenya, and specimens have been found that are between 2 and 2.5 million years old. Some researchers warn, however, that we cannot call *Homo rudolfensis* a bigger version of *H. habilis* as there are distinct differences. *Homo rudolfensis* has a much flatter face from side to side and its cheekbones are further forward, which seems to echo features of *Australopithecines* despite the more modern-looking face. One theory is that these two species were eating different food sources, which explains why they were able to live in the same region without competing for resources.

The situation has become even more complicated with a new synthesis of the early *Homo* fossils published by Susan Anton (New York University), Richard Potts (Smithsonian Institution), and Leslie Aiello (Wenner-Gren Foundation) in *Science* in 2014. They contend that morphologically there are three distinct groups instead of two (Figure 6). They define early *Homo*, which includes the earliest *H. habilis*, starting at 2.35 million years ago; but now, with the latest find, this should be extended to 2.8 million years ago. Then they define two different groups, both appearing at about 2.09 million years ago: one with a relatively tall, flat face, which appears in the fossil record between 2.09 and 1.78 million years ago; and the second with a more primitive face, which appears in the fossil record between 2.09 and 1.44 million years ago. What this new

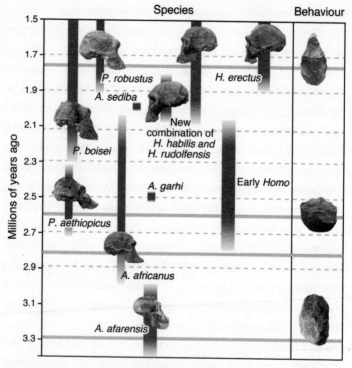

FIGURE 6 New divisions at the start of the *Homo* lineage suggested by Susan Anton and colleagues based on skull morphology, which challenges the original *H. habilis/H. rudolfensis/*early *Homo* split. Figure includes the new stone tools (3.3 Ma) and early *Homo* (2.8 Ma) finds.

synthesis shows is that hominin evolution, rather than being a simple linear transition between clearly defined species, is much more of an evolutionary bush with lots of slightly different morphologically defined species coexisting. We can speculate that these very closely related species were interbreeding, hence why their morphologies are very similar.

Our understanding of early *Homo* may be on the cusp of another revolution. In September 2015, a new species, *Homo naledi*, was announced. It is named after the Rising Star cave where the fossils were first found; this is some thirty miles north-west of Johannesburg in South Africa (Figure 3), and '*naledi*' means 'star' in Sesotho, a local South African language. So far, the team has recovered parts of at least fifteen individuals of the same species, a small fraction of the thousands of fossils believed to remain in the chamber. The team, led by palaeoanthropologist Lee Berger at the University of the Witwatersrand in South Africa, suggests that *naledi* looks like one of the most primitive members of the *Homo* genus. *H. naledi* had a tiny brain, about 500 cm^3, similar to the *Australopithecines* and smaller than *H. habilis*, perched atop a very slender body. *H. naledi* stood approximately 1.5 m (about 5 feet) tall and weighed about 45 kg. *H. naledi*'s teeth and skull are similar to *Homo habilis*. The shoulders, however, are more similar to those of apes. In addition, *H. naledi* had extremely curved fingers—more curved than almost any other species of early hominin—which clearly demonstrates climbing capabilities. This contrasts with the feet of *H. naledi*, which are virtually indistinguishable from those of modern humans. Combined with its long legs, this suggests the species was well suited for long-distance walking. Like any new find, *H. naledi* is very controversial. Some palaeoanthropologists have suggested that it is in fact a small-brained *Homo erectus*, while others dispute whether it is in the *Homo* genus. Regarding the later dispute, the discoverers did not publish their results in the highest impact journals, which is the normal route. Instead they published in a less well-known open-access journal, which they suggest was to enable everyone to have access to the data; others have suggested, more cynically, that this was a way to get the genus name

Homo accepted in the literature. In 2017 dates were obtained from the fossil teeth, sediments encasing the fossils, and overlying flowstone—they were surprisingly young, between 335,000 and 236,000 years ago. No similarly small-brained hominin is known from such a recent date in Africa and we must wonder at the ability of such a hominin to survive for so long in the midst of more advanced members of *Homo*. Some argue it is a small bodied relative of *H. erectus*, others a species that has undergone dwarfism similar to *H. floresiensis*, which is discussed later.

At the same time as early *Homo* was appearing in its various forms, another completely different group of hominins was evolving—the *Paranthropus* group. These hominins had robust dentition and large jaw muscles, and appeared around 2.5 million years ago. They are fascinating as they show us there was a radically different way of being a hominin. The *Paranthropus* genus includes the East African *P. aethiopicus* (2.5 million years ago), *P. boisei* (2.3–1.2 million years ago), and the Southern African *P. robustus* (1.8–1.2 million years ago). All the *Paranthropus* species are thought to have been bipedal, and they had robust skulls to support the large jaws and strong chewing muscles. This adaptation of the skull is thought to have enabled *Paranthropus* to eat hard foods such as nuts, seeds, and tubers. This interpretation is supported by evidence of extreme wear on the enamel of the teeth that have been found. Intriguingly, these species are often found at the same sites as our closer relatives, and it is very likely that if you had visited South Africa between 2 and 1.5 million years ago you could have seen two very different bipeds—one looking vaguely familiar using tools to obtain meat and vegetables, and the other, *P. robustus*, moving across the landscape consuming mainly vegetables, particularly hard nuts and edible roots. Another significant feature of the

Paranthropus genus is the large difference between the size of male and females—an example of what is called sexual dimorphism. *P. boisei* and *P. robustus* males were about twice the size of the females, which is the size difference seen today in gorillas. This ratio is much larger than early *Homo* or modern humans, where males tend to be only slightly larger than females. The fossils of *P. robustus* found at Swartkrans in South Africa also give us a stark reminder of the dangers faced by hominins. Many of the bones show evidence that the individuals were killed by predators. The best example is the skull of a young hominin that has bite marks from a leopard, which had jumped on its back and bitten down on the skull, puncturing both the back of the skull and through the eye sockets.

Stage 4: *Homo erectus*

Arguably the most important episode in hominin evolution occurred in East Africa around 1.9–1.8 million years ago, when hominin diversity reached its highest level, with species of the *Australopithecus, Paranthropus,* and *Homo* genera all coexisting alongside each other. At the same time, the most important leap forward in human evolution appears to have occurred with the appearance of *Homo erectus* (understood in the broadest sense), which is associated with sweeping changes in brain size, life history, and body size and shape. Until very recently it was assumed that *Homo erectus* was the first hominin to leave Africa—but this has been called into question by a beautiful set of stone tools found in sediment from the Chinese Loess Plateau, near Gongwangling in Lantian county dated to 2.1 million years ago. This raises the question whether *Homo erectus* evolved earlier than we thought or another hominin species made it all the way to China. I say *Homo erectus* as understood in the broadest sense because African specimens, and sometimes those from Dmanisi in Georgia, are described as *Homo ergaster*. The name 'ergaster' means 'workman' in Greek and was used as the African specimens have all been found with stone tools. Moreover, *H. ergaster* does not have all the features that are found in Asian populations of *H. erectus*. To help the reader I will, from now on, refer to this group of species simply as *H. erectus* and hope my palaeoanthropologist colleagues can forgive me. Post-cranially, *H. erectus* is very similar to modern humans. Its brain size was about two-thirds the size of a modern human which is much larger than earlier hominins. Early representatives of *Homo erectus* in Africa had a brain that was over 80 per cent larger than *Australopithecus afarensis* and 40 per cent larger than *Homo habilis*. In contrast, from the appearance of the early *Australopithecines* until the appearance of the first member of the genus *Homo*, there was remarkably little change in hominin brain size (Figure 7). In fact, many of us would argue that the *Homo* genus should really start with *Homo erectus*, given its similarity to modern humans and the significant differences with earlier *Homo* specimens.

For me, the emergence of *H. erectus* in East Africa represents a fundamental turning point in hominin evolution. The dramatic increase in brain size was also accompanied by major changes in life history and body morphology. First the pelvis morphology changed to allow the birth of a larger head associated with a bigger brain. There is also beautiful evidence from growth lines in the fossilized teeth that *Homo erectus* was the first hominin to have a delayed growth period during childhood. Modern human children grow very differently to the offspring of other apes. A male chimpanzee and male human both end up with the same body weight but they have quite different growth patterns. At age one

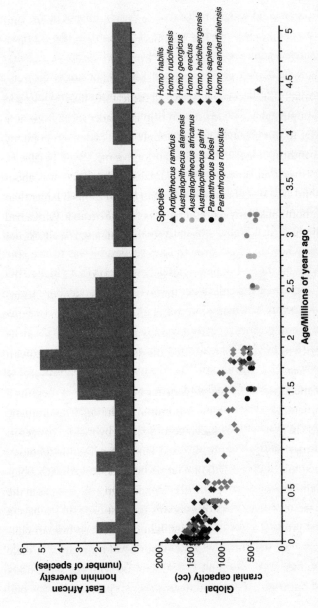

FIGURE 7 Records of hominin brain capacity and East African hominin diversity over the past five million years.

the human child weighs twice that of the chimp, but at age eight the chimpanzee is twice the weight of the human. The chimpanzee gains its adult weight at age twelve—a whole six years before the human. A male gorilla is also a fast and consistent grower; a 150 kg male gorilla weighs 50 kg by its fifth birthday and 120 kg by its tenth birthday. This is because humans have a growth pause or plateau, extending childhood and delaying full adult development, and this pattern began with *H. erectus*. The body size of *H. erectus* is larger than that of earlier hominins and comparable to modern humans. This is well illustrated by the *H. erectus* skeleton known as the Nariokotome Boy, found in 1984 by a team led by Richard Leakey and Alan Walker at Lake Turkana. Hip analysis shows that he was a boy, and teeth growth lines suggest he was twelve years old when he died—but he was already 160 cm (5 foot 3 inches) tall. Dan Lieberman and colleagues at Harvard University have shown that *H. erectus* had many key adaptations required for long-distance running. More recently it has also been shown that the shape of the shoulder in *H. erectus* would have allowed the throwing of projectiles. *H. erectus* also produced a much more sophisticated set of stone tools than previous hominins, referred to as Acheulean tools. It has also been argued that *H. erectus* had learnt to control fire because, as Richard Wrangham at Harvard University eloquently argues, it is difficult to see how they could have maintained such a large, energy-intensive brain with such a small gut without access to cooked meat. However, this runs contrary to the prevailing view of most palaeoanthropologists that the control of fire and regular cooking came much later in human evolution. But if this is correct, how did *H. erectus* get enough calories to fuel its bigger brain? Katherine Zink, also at Harvard University, and Daniel Lieberman may have found a solution. They experimented with processing food with stone tools that would have been available to *H. erectus*. By slicing meat and pounding root vegetables and nuts they were able to improve the ability to chew by at least 40 per cent. There was also a decrease in the masticatory force needed, which corresponds to the observed reduction in jaw size and strength in *H. erectus*. The control of fire and regular cooking was, as Rick Potts at the Smithsonian suggests, essential for the next significant increase in brain size between 700,000 and 600,000 years ago with the appearance of *H. heidelbergensis* and then Neanderthals and *H. sapiens*.

So *H. erectus* was smart, with a large and flexible stone toolkit, able to run long distances, throw projectile weapons, and process food. If we are looking for the most successful hominin species in history then the answer would have to be *H. erectus*, as it survived from 1.8 million years ago to 200,000 years ago and colonized much of Africa and Eurasia. In fact, *H. erectus* may have survived even longer. In 2003, partial skeletons of nine individual hominins were recovered from a cave on the island of Flores in Indonesia. These individuals stood about 1.1 m tall (3.5 feet). Not only were they small but they also had a correspondingly small brain, about 380 cm^3, somewhere between chimpanzees and *Australopithecines*. The new species was called *Homo floresiensis* or Flores Man, and nicknamed 'hobbit' or 'Flo'. The earliest evidence for *H. floresiensis* are a few teeth and a mandible found at Mata Menge in the Soa Basin and dated to 700,000 years ago. The partial skeletons found at Liang Bua are much younger with stratigraphic evidence from the cave and dates of associated stone artefacts suggesting an age of between 100,000 and 50,000 years ago. One suggestion is that, due to the similarity in morphology, *H. floresiensis* is in fact a population of *H. erectus* that had undergone island dwarfism: species that colonize isolated islands are often found to decrease in size over time. Intriguingly, stories abound that *H. floresiensis* may have survived after 50,000 years ago in other parts of Flores, as the Nage people of the island tell stories of the *Ebu Gogo*—small, hairy, language-poor cave dwellers, which seem to be very similar to *H. floresiensis*. It has been suggested that the *Ebu Gogo* were present at the time of the arrival of the first Portuguese ships during the sixteenth century, and there are even claims they existed as recently as the late nineteenth century. But then again other cultures around the world have their own stories of similar human-like monsters hiding in the forests or the mountains such as trolls, dwarfs, Pukwudgie, Bigfoot, Yeren, and the Yeti.

Stage 5: The journey towards *Homo sapiens*

With the appearance of *H. erectus*, brain size increased significantly and continued to increase over the following half a million years (Figure 7). Then, a million years after *H. erectus* first occurs in the fossil record, a new species, *Homo heidelbergensis*, appears, with the first-known fossil found in Ethiopia dated to 700,000–600,000 years ago. The name *Homo heidelbergensis* has been around a long time, as it was first coined for a jawbone found near Heidelberg, Germany, in 1907. The present concept of the species *H. heidelbergensis* is, however, new and it is understood to represent a descendant of *H. erectus* in Africa which spread out into Europe. *H. heidelbergensis* may even have been the common ancestor of Neanderthals and *Homo sapiens*. *H. heidelbergensis* specimens are recognizable from very human-like but massive robust skulls. Brain capacity ranges from 1,100–1,400 cm³, which is three times larger than an *Australopithecus* brain and, at the top end, overlaps the range of modern humans. Early fossils of *H. heidelbergensis* found in Greece, Spain, Ethiopia, and Zambia look very similar, and this is why the species is thought of as both an African and a European species. Later European populations begin to look different, and this is why it is thought they may have evolved into Neanderthals. *H. heidelbergensis* even made it to Britain—a population which the palaeoanthropologist Chris Stringer lovingly refers to as 'Homo Britannicus'.

As you have probably become aware through reading this chapter, nothing is simple when it comes to naming hominin fossils, and there are two other species names you will come across. The first is *Homo rhodesiensis*, the original species name given to the Zambia specimen which most now consider a clear example of *H. heidelbergensis*. The other is *Homo antecessor*, which is based on the oldest known hominin fossils in Europe found in the mountainous northern Spanish site of Atapuerca. This amazing site is famous for its incredible collection of *H. heidelbergensis* fossils dated to more than 400,000 years ago. But much older hominin fossils have also been found there, dating from 1.2 million to 800,000 years ago. Eighty bones from six individuals have been found, and because of the differences with *H. heidelbergensis* they have been given the species name *H. antecessor*. *H. antecessor* was about 1.6–1.8 m (5.5–6 feet) tall and its brain size was roughly 1,000–1,150 cm³, smaller than most *H. heidelbergensis* fossils.

Between 700,000 and 300,000 years ago there is evidence for only one hominin species in Africa and Europe, *H. heidelbergensis*. Fossilized wooden spears that are 400,000 years old have been found in Germany, and 500,000-year-old hafted stone points used for hunting have been discovered in South Africa, suggesting that *H. heidelbergensis* was an active hunter. *H. heidelbergensis*, like *H. erectus* before it, migrated out of Africa into both Europe and Asia. Though the focus of this book is on our African origins it is worth noting that it is currently thought that European populations of *H. heidelbergensis* gave rise to *Homo neanderthalensis* around 400,000 years ago; Asian *H. heidelbergensis* were ancestral to *Homo denisovan* around 600,000 years ago; and African *H. heidelbergensis* were the ancestors of *H. sapiens* around 200,000 years ago.

The last step on the road to modern humans is the appearance of *Homo sapiens*. An archaic form of *H. sapiens* has been found in Morocco on the Atlantic coast dated to 315,000 years ago. The first true *H. sapiens* specimens, however, are from Omo in south-west Ethiopia, dated to about 195,000 years ago. It is generally accepted that *H. sapiens* evolved in East Africa and then dispersed out of Africa in a similar way to *H. erectus* and *H. heidelbergensis*. But the intriguing point is that, for about 150,000 years, *H. sapiens* behaved in a very similar way to its ancestors. It is only after about 100,000 years ago that there is evidence for abstract thought and the use of symbolism to express cultural creativity. From 50,000 years ago we find spectacular cave paintings in France, Spain, and Indonesia and 'Venus' figurines in Austria, France, and the Czech Republic. In Chapter 8, I examine what kind of revolution may have occurred 50,000 years ago to set in train the cumulative culture which continues today and is one of the defining features of modern humans. Some researchers have suggested that we should perhaps use subspecies to define these changes. So *Homo sapiens idaltu* has been suggested for the more robust primitive form of *H. sapiens* found at sites including Herto in Ethiopia (dated at 160,000 years old) and Skhul in Israel (dated at 90,000 years old), while the terms *Homo sapiens sapiens*, or anatomically modern

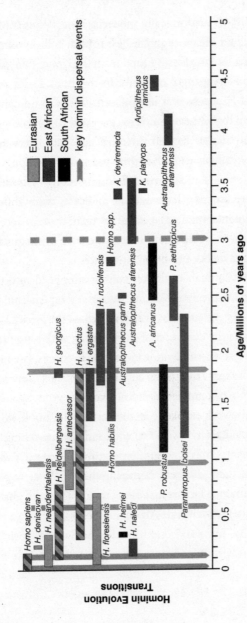

FIGURE 8 Summary of the hominin species covering the past five million years and major dispersal events. The dotted line shows the first dispersal from East Africa into Southern Africa and the solid lines represent dispersals of hominins out of Africa.

humans (AMH), or anatomically modern *Homo sapiens* (AMHS), should be used for the more gracile, less robust, and almost completely modern-looking fossils found from the Upper Palaeolithic period (~50,000 years ago) onwards. In this book I stick to the convention of *H. sapiens* and use the term modern humans for those *H. sapiens* that appeared after 50,000 years ago.

I hope this chapter has provided an adequate overview or framework of human evolution over the past 7 million years. With such a small and fragmented fossil record of hominins available, there are, as I have tried to indicate, many different opinions and controversies. Figure 8 summarizes what we currently know about the first and last appearance of the different hominin species mentioned here.

For me, early human evolution can be simplified into three stages. The first is the evolution and success of bipedalism, with the spread of the *Australopithecus* species across the African continent. Second is the evolution of *Homo erectus*, with a significant jump in brain size and body morphology to become close to that of modern humans. Third is the appearance of *Homo sapiens* and the acceleration of cumulative culture 50,000 years ago. Now, with this framework in place, in the rest of the book we will explore how plate tectonics, global and local climate change, and celestial mechanics may have influenced the appearance and disappearance of these different hominin species. What were the wider environmental factors that drove the evolution of humans, and why did they arise in East Africa?

3

Tectonics and Climate

To understand long-term changes in global climate and their effects on evolution we need to understand the underlying driving force of tectonics. The Earth's climate is simply a function of how the Sun's energy falling on the Earth is redistributed around the globe. This heat redistribution is influenced by the position of the continents and the oceans, which is controlled by plate tectonics. This is why 100 million years ago the Earth was much warmer and more humid than today and dinosaurs were happily living in Antarctica.

Tectonics has two main effects on climate. First, there are the direct effects, which include the creation of mountains and plateaux, which block atmospheric circulation and change the hydrological cycle. They also include the movement of continents that changes the shape of the oceans, creating ocean gateways. Second, there are indirect effects, which regulate atmospheric greenhouse gases through subduction, volcanism, and the chemical weathering of rocks. Understanding how tectonics influences climate is essential if we are to understand human evolution, because without the uplift of the Tibetan Plateau and the formation of the East African Rift Valley hominin evolution would not have occurred.

In this chapter I break down the influences into horizontal tectonics, which looks at what happens if you simply move the continental plates around the globe; and vertical tectonics, which looks at what happens if you create mountains or plateaux. Finally, we will consider how these changes influenced hominin evolution in Africa.

Horizontal tectonics

Latitudinal continents

The north–south position of the continents has a huge effect on the thermal gradient between the poles and the equator. Geologists have run simple climate models to look at this effect (Figure 9). If you put all the continents around the equator—the so-called 'tropical ring world'—the temperature gradient between the poles and the equator is about 30°C (Figure 9). When the poles are covered by ocean, they cannot go below freezing, due to a trick of both the atmosphere and the oceans. A fundamental aspect of climate is that hot air rises and cold air sinks. At the poles it is cold so the air sinks, and as it hits the ground it spreads outwards away from the pole. When sea water at the pole freezes it forms sea ice, and this ice is then blown away from the pole towards warmer water, where it melts. This maintains the balance and prevents the temperature at the poles going below zero. However, as soon as you place land at the poles, or even around the poles, ice can form permanently. If you have a landmass like Antarctica over a pole with ice, then ground temperatures drop below −35°C, which creates an equator–pole

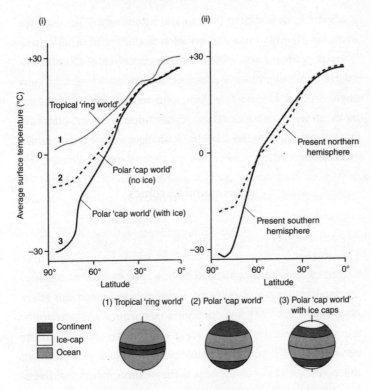

FIGURE 9 Illustration of the huge effect the latitudinal location of continents have on the equator–pole temperature gradient. The present-day northern and southern hemisphere temperature gradients are given as a comparison.

temperature gradient of over 70°C (Figure 9)—exactly what we have today.

In contrast, if you consider the present northern hemisphere, the continents are not on the pole, but surround it. So instead of one huge ice sheet, as we have in Antarctica, there is a smaller one on Greenland, and the continents act like a fence and keep all the sea ice in the Arctic Ocean. So the equator–pole temperature

43

gradient of the northern hemisphere is somewhere between the extremes of Antarctica and an ice-free continent—about 50°C. The size of the equator–pole temperature gradient is a fundamental driver of our climate, because ocean and atmospheric circulation is in part driven by the need to move heat from the equator to the poles. This temperature gradient defines what sort of climate the world will have. A cold Earth has an extreme equator–pole temperature gradient and thus a very dynamic climate. This is why we have strong hurricanes and winter storms as the climate system tries to pump heat away from the hot tropics towards the cold poles.

Longitude continents

Ocean circulation is controlled by how the oceans are contained by the continents around them. If there are no continents in the way then oceans will just continue to circulate round and round the globe. However, when an ocean current encounters a continent it is deflected both north and south. If we look at the modern configuration of the continents, there are three main longitudinal continents: (1) the Americas; (2) Europe down to Southern Africa; and (3) North East Asia down to Australasia. A hundred million years ago the continents would have been still recognizable, but in slightly different positions. The two striking features are that, first, there was an ocean across the whole of the tropics through the Tethyan Sea and the Deep Central American passage, and second, there was no ocean circulating around Antarctica. These changes have had huge effects on the circulation of the surface ocean and thus deep ocean circulation and global climate.

The circulation of the surface ocean is important in transporting heat around the globe. The circulation of the surface oceans is primarily controlled by the prevailing winds, because as the wind blows on the surface water the friction allows the energy to be transferred from winds to the surface water that leads to the major currents. The wind energy is transferred to greater depths in the water column turbulence, which allows wind-driven currents to be very deep. The rotation of the Earth adds an extra twist, because of the Coriolis effect, which means in the northern hemisphere the ocean currents will move to the right of the prevailing wind direction while in the southern hemisphere they will move to the left.

In the tropics the prevailing wind direction is from both the north-east and south-east, and hence the surface ocean currents move westward. In the subtropics the winds come from the west and thus surface ocean currents are pushed eastward. In the high latitudes the winds are again from the west and so the surface ocean is pushed eastward. These currents will flow as far as they can until they hit an obstacle such as a continent, when the currents are deflected both north and south along the continental margin. These giant currents join up within each ocean basin and form gyres. These large systems of rotating ocean currents are found in the present-day North and South Atlantic Oceans, North and South Pacific Oceans, and the Indian Ocean.

By understanding the modern circulation of the oceans the effects of ocean gateways on ocean circulation can be investigated by modelling three very different scenarios. The first is a simple double-slice world with longitudinal continents on either side (Figure 10a). Ocean currents are primarily driven by the wind so in the tropics and high latitudes ocean currents are pushed westward, while in the mid-latitudes they are pushed eastwards. This produces the classic two-gyre solution

a)

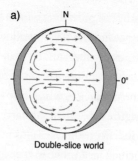

Double-slice world

b)

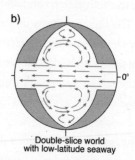

Double-slice world
with low-latitude seaway

c)

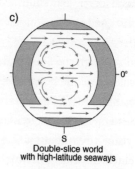

Double-slice world
with high-latitude seaways

FIGURE 10 Illustration of the significant alteration of ocean currents with different ocean gateways within longitudinal continents.

in both hemispheres. Today, both the North Pacific Ocean and the North Atlantic Ocean have this type of circulation. The second scenario is a double-sliced world with a low-latitude seaway. This produces a large tropical ocean circulating continually westward around the world. There are then two smaller gyres in each hemisphere (Figure 10b). This is the circulation seen during the Cretaceous Period, with the two gyres in each hemisphere occurring in the Pacific Ocean. The remnants of the Cretaceous low-latitude seaway, the Tethys Sea, can still be seen today in the Mediterranean Sea. The linkage between the Tethys Sea and the Indian Ocean finally closed between 11 and 7 million years ago. This coincides with the first sand dune deposits found in the Sahara desert dated at 7 million years ago. It was originally assumed that the Sahara desert had come into existence with the start of the great ice ages at about 3 million years ago. Climate modelling work now suggests that the shrinking of the Tethys Sea weakened the African summer monsoon, allowing desert conditions to expand across North Africa and created a split between African and Asian flora and fauna. This work also suggests that the final closing of the Tethys Sea enhanced the sensitivity of the African monsoon to orbital forcing (see Chapter 6), which later drove the expansions and contractions of the Sahara desert. About 5.5 million years ago tectonics closed off the link between the Mediterranean Sea and the North Atlantic Ocean, and over a period of 200,000 years it repeatedly dried out and had a profound effect on the climate of North Africa.

The third scenario is a double-sliced world with high-latitude seaways. This produces strong circumpolar ocean currents and a single tropical gyre in each hemisphere (Figure 10c). Today, the southern hemisphere resembles this scenario, with a circumpolar current around Antarctica. The Southern Ocean thus acts like a giant heat extractor and was instrumental in the huge build-up of ice on Antarctica.

Deep ocean circulation

Deep ocean circulation is also an important consideration, as it influences the circulation of the surface ocean and the distribution of heat between the hemispheres. The presence or absence of ocean gateways has a profound effect on deep ocean circulation. For example, our modern-day North Atlantic Deep Water (NADW), which helps pull the Gulf Stream northwards, maintaining a mild European climate, may be only 4–5 million years old—about the same time *Australopithecines* were spreading across Africa. If we run computer simulations of ocean circulation with and without the Drake Passage and the Panama ocean gateway, only the modern-day combination produces significant NADW. Hence, our modern-day deep ocean circulation is due to an open Drake Passage from about 25 million years ago and the closure of the Panama ocean gateway from about 4 million years onwards (Figure 11). It is all due to salt. Because of the greater effect of evaporation in the North Atlantic region, the North Atlantic Ocean is saltier than the Pacific Ocean. NADW forms today when the warm salty water from the Caribbean travels across the Atlantic Ocean and cools down. The high salt load and colder temperature both increase the density of the water, so it is able to sink north of Iceland. When the Panama passage was open, fresher Pacific Ocean water leaked in and reduced the overall salt content of the North Atlantic Ocean. The surface water, even when cooled, was thus not dense enough to sink, and so not as much NADW was

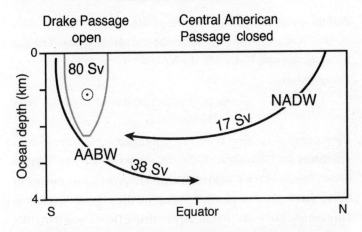

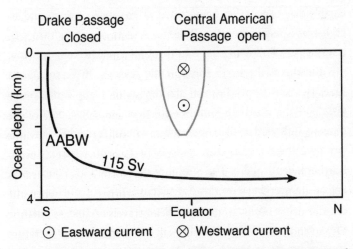

FIGURE 11 The effect of changing ocean gateways on the location and intensity of deep ocean circulation. NADW = North Atlantic Deep Water, AABW = Antarctic Bottom Water, and Sv = Sverdrup, that is, 1 million cubic metres per second.

formed compared to today. So one of the fundamental elements of our modern climate system—the competition between the Antarctic Bottom Water and the NADW—turns out to be a very young feature.

Monsoons

The position of the continents in the tropics is also very important as they can create monsoonal systems. The name monsoon comes from the Arabic word *mausim*, which means season, as most of the rains that fall in South East Asia occur during the northern hemisphere summer. In the tropics the Sun's energy is most intense when it is overhead. The seasons are caused by the Earth's axis of rotation being tilted at an angle of 23.5°, and so each hemisphere receives different amounts of energy throughout the year. Following the Sun through a year we can see how this tilt affects the Earth through the seasons. If we start at 21 June, the Sun is overhead at midday at the Tropic of Cancer (23.4°N)—the northern summer solstice. The angle of the Sun moves southward until 21 September, when it is overhead at midday over the equator—the equinox or autumn equinox in the northern hemisphere. The Sun appears to continue southward, and on 21 December is overhead at midday at the Tropic of Capricorn (23.4°S)—the southern summer solstice. The Sun then appears to move northward until it is directly overhead at midday at the equator on the 21 March—the equinox or spring equinox in the northern hemisphere. And so the cycle continues.

During the summer in each hemisphere the continents receive the most solar energy and heat up. This causes the air above the land to warm and rise, creating an area of low pressure beneath it. This low pressure sucks air from the surrounding area and, if this is from an adjacent ocean, very moist air is brought over the continent. As this moist air is warmed it rises, and as it gets higher it cools, the water condenses, and the extremely heavy rains we refer to as monsoons are created. In Africa there are two monsoonal systems. The first is in West Africa: during the northern hemisphere summer, North Africa heats up and draws in moist air from the Atlantic Ocean, creating the Congo monsoonal system. This system is so strong that Atlantic Ocean-derived moisture is pulled across the whole continent into East Africa. The second monsoonal system is in East Africa. Again, during northern hemisphere summer North Africa heats, up but in East Africa the moist air is drawn from the Indian Ocean. However, as we will see later, because of the East African Rift Valley mountains and the Tibetan Plateau, much of this moist air is deflected and joins the intense South East Asian monsoonal system.

Vertical tectonics

As the tectonic plates bearing the continents move around the surface of the Earth they frequently collide. When this happens the land is pushed upwards. In some cases chains of mountains are formed; or, when whole regions are uplifted, plateaux are formed. These have a profound effect on the climate system. One of these effects is a rain shadow, which is a dry area on the leeward side of a mountain system. There is usually a corresponding area of increased precipitation on the forward side. As a weather system at ground level moves towards a mountain or plateau it is usually relatively warm and moist (Figure 12). As the

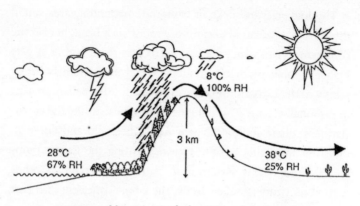

FIGURE 12 Mountain rain shadow (RH = relative humidity).

air encounters the mountain it is forced to move up and over it. Because atmospheric pressure decreases with increasing altitude the air has to expand, and as it does it cools down. Cool air can hold less moisture than warm air so the relative humidity rapidly rises until it hits 100 per cent, and strong rainfall occurs. As the air descends on the other side of the mountain, atmospheric pressure increases, the air temperature rises, and the relative humidity drops very low, because little or no moisture is left in the air. Hence, on the descending side there is a rain-shadow area as there is no moisture left to form rain, and this can lead to the creation of a desert. This simple process can control the wetness or dryness of whole continents. Figure 13 shows the effect of mountains occurring on the western or eastern boundary of a continent.

There are three main rainfall belts in the world: one in the tropics and one each in the mid-latitudes of the two hemispheres. Air in the tropics moves from east to west, while in the mid-latitudes it moves west to east. Having mountains on the western side

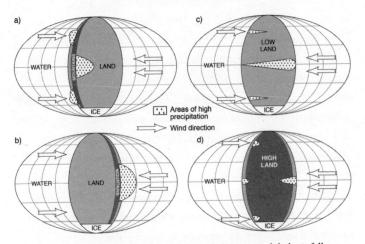

FIGURE 13 The effects of mountains and plateaus on global rainfall.

produces more rainfall on land and a wetter continent overall. By coincidence, at the moment we have western mountain ranges running down the west coast of North America—the Rockies—and the west coast of South America—the Andes. These mountains not only produce significant wet areas but also famous deserts like the Atacama in Chile and Death Valley in the United States—two of the driest deserts on Earth. In contrast, when the East African Rift Valley was formed it created mountain ranges up to 4 km high on the eastern boundary of the Africa continent—hence a lot of the moisture is prevented from falling over East Africa and is lost to the Indian subcontinent.

Atmospheric barriers

When huge mountains or plateaus are thrust high up in the sky they interfere with the circulation of the atmosphere. Not only do they force air up and over them, but in many cases they deflect the weather system around them. This effect is compounded as up-lifted areas also warm up in summer and cool down in winter more than the surrounding lowlands. Figure 14 shows that if all the continents in the northern hemisphere were flat, then the major circulation of the atmosphere would be nearly circular, with maybe a slight deflection due to the difference between land and the oceans. However, if you put the two modern plateaux in place—i.e. the uplifted regions of the Tibetan–Himalayan and Sierran–Coloradan complexes—then there are huge changes in circulation. Both these plateaux are massive. The Tibetan Plateau is the world's highest and largest, with an area of 2.5 million km², which is about four times the size of France. The Colorado Plateau covers an area of 337,000 km² and is joined to numerous other plateaux, which together make up the Sierran–Coloradan uplift complex.

In the northern hemisphere winter these highlands are much colder than the surrounding areas, creating a high-pressure sys-tem and out-blowing anti-cyclonic circulation (see Figure 14a). This deflects Arctic air northwards and keeps the middle of the Asian and North American continents warmer than they would otherwise be. In the northern hemisphere summer these two major plateaux heat up more than the surrounding areas and thus the air above them rises, creating an intense low-pressure zone (Figure 14b). This sucks in surrounding air, producing a cyclonic circulation-deflecting weather system much further north and south. The summer cyclonic circulation around the Tibetan–Himalayan plateau also greatly intensifies the South East Monsoonal system. Because part of the air that is pulled towards the Himalayas comes from the Indian Ocean it brings with it a lot of moisture, part of which is taken away from East Africa. The

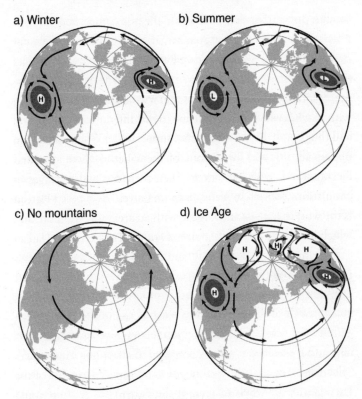

a) Winter

b) Summer

c) No mountains

d) Ice Age

FIGURE 14 The effects of plateaus and ice sheet on atmospheric circulation.

South East Monsoonal rainfall is essential for the well-being of two-fifths of the world's population.

Volcanic eruptions

Plate tectonics drives the formation of volcanoes, which have an important influence on climate through the introduction of gases and dust into the atmosphere. Normal-sized volcanoes inject sulphur dioxide, carbon dioxide, and dust into the troposphere and can have a considerable effect on our weather. The explosion of Krakatoa in 1883 killed over 36,000 people and is considered to be the loudest sound ever heard in modern history, with reports of it being heard nearly 3,000 miles away. The eruption was equivalent to 200 megatons of TNT, which is about 13,000 times the nuclear yield of the Little Boy bomb that devastated Hiroshima during World War II. The sulphur dioxide and dust injected into the atmosphere increased the amount of sunlight reflected back into space, and average global temperatures fell by as much as 1.2°C in the year following the eruption. Weather patterns continued to be chaotic for years and temperatures did not return to normal until 1888.

The effect of Krakatoa on climate was short term and transient. This is because the sulphur dioxide and dust were injected relatively low in the atmosphere and the amount of water also injected meant much of the material was washed out of the atmosphere within a few years. Volcanic eruptions experienced during human history have been very small compared to eruptions from supervolcanoes. These are thousands of times larger than Krakatoa. They can occur when magma in the Earth rises into the crust from a hotspot but is unable to break through the crust. Pressure builds in a large and growing magma pool until the crust is unable to contain the pressure. Supervolcanoes can also form at convergent plate boundaries—for example, Toba, which last erupted about 74,000 years ago and ejected about 2,800 cubic kilometres of material into the atmosphere. And they can form in continental hotspot locations—for example, Yellowstone, which last erupted 2.1 million years ago and ejected 2,500 cubic kilometres of material. Because of the scale of these events the sulphur dioxide and dust are injected much higher in the atmosphere and therefore the effects on global climate can last much longer. Modelling work by the UK Meteorological Office suggests that a single tropical supervolcano eruption would cause a drop of global temperatures of at least 6°C—up to 15°C in the tropics—for a minimum of three years. Then, over a couple of decades, the climate would slowly come back to within 1°C of normal. But imagine what would happen if you had a succession of supervolcano eruptions. This is exactly what happened about 66 million years ago in India; making the global climate see-saw between warm and cold conditions on a decadal timescale.

The Deccan Traps are massive lava flows which began forming about 66.25 million years ago, at the end of the Cretaceous Period. It is estimated that when formed over a 30,000-year period the lavas covered 1.5 million km^2, which is approximately half the size of modern India. Due to erosion and plate tectonics the Deccan Traps have been reduced to their current size of around 500,000 km^2. The release of volcanic gases, particularly sulphur dioxide, during these frequent supervolcanic eruptions, is thought to have dropped global temperatures by over 2°C. It has been speculated that these frequent recurring cooling events may have played a role in the Cretaceous–Paleogene extinction event, when non-avian dinosaurs were made extinct. However, the current consensus among scientists is that this extinction was triggered by the Chicxulub meteorite impact event in Central America around 66 million years ago, as this would have produced a sunlight-blocking dust cloud that killed much of the plant life and reduced global temperature, during what is now called an impact winter. However, recent work by Gerta Keller at Princeton and others suggests that the extinction may have been caused by both the volcanism and the impact event. This extinction event was essential in the story of human evolution as it ended the 130-million-year domination of dinosaurs and allowed the evolution and proliferation of mammals, in particular the appearance of the first ancestors of primates.

Icehouse and greenhouse worlds

Plate tectonics drives the slow shift of the continents across the globe, combining them into supercontinents and then back into fragmented continents again. The supercontinent Rodinia formed about 1.1 billion years ago and broke up roughly 750 million years ago. Another supercontinent, called Pannotia or the Vendian Supercontinent, formed about 600 million years ago, but only lasted about 60 million years. One of the fragments included large parts of the continents we now find in the southern hemisphere. Plate tectonics brought the fragments of Rodinia back together in a different configuration about 300 million years ago, forming the best-known supercontinent, Pangaea. Pangaea subsequently broke up into the northern and southern supercontinents of Laurasia and Gondwana, about 200 million years ago. Both of these supercontinents have continued to fragment over the past 100 million years. Icehouse climates form when the continents are moving together. The sea level is low due to lack of sea-floor production. The climate becomes cooler and more arid because of the reduction in rainfall due to the strong rain-shadow effect of large superplateaux. On the other hand, greenhouse climates are created as the continents are dispersed, with sea levels high due to the high level of sea-floor spreading. There needs to be a relatively large amount of carbon dioxide in the atmosphere, maybe over three times current levels, due to production at oceanic rifting zones. This produces a warm, humid climate.

The formation and break-up of these supercontinents had a huge effect on evolution (Figure 15). Supercontinents are extremely bad for life and are often associated with mass extinctions. First, the interior of the continent is very dry and global climate is usually cold. Second, there is a massive reduction in the amount of shelf seas. It is not surprising that the explosion of complex animals occurred during the Cambrian Period beginning 542 million years ago, following the break-up of the Vendian supercontinent, as the amount of shallow seas in which complex life first evolved increased exponentially as the supercontinent fragmented. The new continental fragments became less and less arid, and thus increasingly hospitable for life. Mass extinctions are correlated with the formation of supercontinents. For example, it is estimated that up to 96 per cent of all marine species and 70 per cent of terrestrial vertebrate species became extinct during the Permian–Triassic extinction event 250 million years ago—nicknamed the 'mother of all mass extinctions' (Figure 15). This mass extinction ended the dominance of the 'proto-mammals' and led to the evolution of dinosaurs and mammals though, for the next 170 million years, the dinosaurs reigned supreme.

Understanding how tectonics affects global climate suggests the influence of the formation and break-up of supercontinents on the development of complex animals 542 million years ago, and the evolution of mammals at 225 million years ago. A combination of supervolcanoes erupting in India and a massive meteorite impact around 66 million years ago ended the dominance of the dinosaurs and enabled the mammals to diversify, including the appearance of the early ancestors of primates. Tectonics also influences the location of the continents and thus where monsoonal systems form. The uplift of mountain ranges

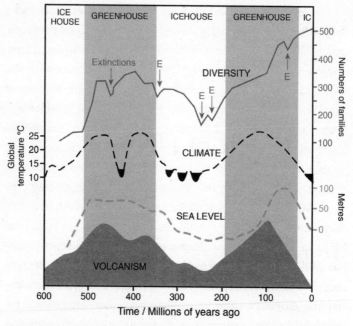

FIGURE 15 Long-term links between tectonics, sea level, climate, biodiversity, and extinctions.

and plateaus influences rainfall and temperature patterns. As we will see in Chapter 4, the uplift of the Tibetan Plateau that started 20 million years ago intensified the South East Asian monsoon, drawing moisture away from East Africa. This was further exacerbated by the formation of the East African Rift Valley mountains, which deflected much of the moisture-laden air towards the Indian subcontinent. This aridification of East Africa and the creating of a mosaic of different vegetation types seems to be one of the key driving forces of early human evolution.

4

Cradle of Humanity

Tectonics can cause significant changes in climate, hydrology, and vegetation cover, both globally and regionally. In this chapter we look at the recent history of tectonics in East Africa and how this has changed the local climate and vegetation. The long-term climatic change in East Africa was controlled by the progressive formation of the East African Rift Valley leading to increased aridity, a fragmentation of the vegetation, and the development of many lake basins. This made the region highly responsive to climate change, and created a dynamic environment in which our ancestors lived and evolved.

Formation of the East Africa Rift System

East Africa is an example of an area that has undergone the process of active rifting. This is initiated by significant uplift due to a magmatic hot spot beneath the crust. Hot spots are caused by plumes of hot magma, which rise from deep in the mantle. Where a hotspot occurs beneath a continent, it heats the continental

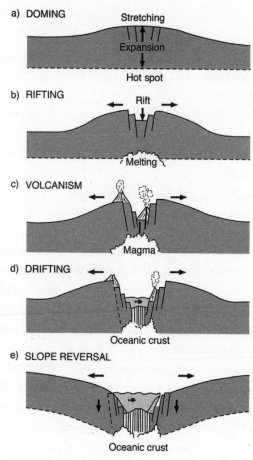

FIGURE 16 Tectonic rifting and ocean formation.

crust, which expands and stretches, and this bulging of the crust leads to the formation of a high plateau (Figure 16). Finally, the expansion becomes so great that the rocks fracture, creating faults on either side of the plateau. These faults allow the central part of the plateau to slip down, creating a giant hanging valley and leaving large mountain ranges on either side. The final stage of rifting occurs when the bottom of the hanging valley drops so low the continental crust becomes very thin and upwelling basaltic magma initiates the formation of oceanic crust. At this stage, because of the subsidence of the relatively dense oceanic crust, the rift valley becomes flooded with seawater, and this can be the start of a new ocean. The Atlantic Ocean was formed in this way 130 million years ago, when the Americas and Africa started to move apart.

The East African Rift System is one of the most extensive geological features on the Earth's surface. It runs north–south for around 4,500 km from Syria through East Africa down to Mozambique. We think it was formed by a hot spot that was centred on northern Ethiopia, the Horn of Africa, and the southern Arabian region. As with most rift processes, the hot spot created a triple junction, with the fracturing of the rocks occurring in three different directions (Figure 17). In this case there is fracturing south along the East African Rift Valley; north-west, forming the Red Sea; and north-east, creating the Gulf of Aden running into the northern Indian Ocean. Volcanism in East Africa may have started in Ethiopia as early as 45–33 million years ago, while initial uplift may have occurred between 38–35 million years ago. There is evidence for volcanism as early as 33 million years ago in northern Kenya, but magmatic activity in the central and southern rift segments in Kenya and Tanzania did not start until between 15 and 8 million years ago. The high relief of the East African Plateau developed between 18 and 13 million years ago, while major faulting in Ethiopia occurred between 20 and 14 million years ago and was followed by the development of east-dipping faults in northern Kenya between 12 and 7 million years ago.

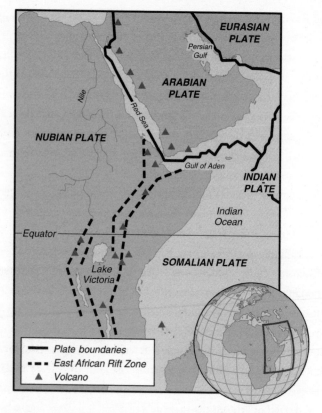

FIGURE 17 Map of plate boundaries at the African Rift Valley triple junction.

Subsequent rift expansion has led to distinctive sets of faults, creating a landscape dominated by grabens and half-grabens.

A graben is the geological name of the part of the crust that
drops when there are vertically orientated faults (Figure 18). A half-
graben arises when there is only one set of faults, creating very
distinctive highlands or horsts, with one side having a very steep
slope (the 'footwall') and the other side (the 'hanging wall') a

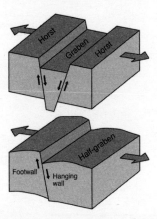

FIGURE 18 Explanation of horst-graben and half-graben geological formations.

gently sloping gradient. In Kenya, half-graben morphologies developed between 9 and 6 million years ago, whereas the full graben
morphology was established between 5.5 and 3.7 million years ago.
By 2.6 million years ago, the Kenya Rift was further fragmented by
new wet-dipping faults, which dropped part of the region down
to a 40-metre-wide inner rift, leaving a 30-kilometre-wide intrarift
plateau: the Kinangop Plateau.

This rifting process has created a unique landscape (Figure 19).
In the centre of the Rift Valley system there is a large hanging
valley about 1,000 m above sea level; to either side are asymmetric mountain ranges with very steep sides facing the valley, and
then gentle slopes down the other side. It is amazing to fly
across this landscape when getting to and from fieldwork sites
in the Rift Valley. You get a real sense of the saw-tooth shape of
the Rift shoulder mountains which steadily get higher and
higher the further you get from the valley. Add to this a smaller
rifted valley within the central Rift Valley and you get a sense

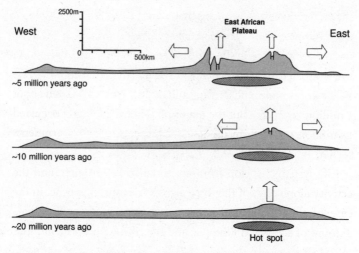

FIGURE 19 Tectonic development of the East African Rift Valley system.

of how complex and fragmented this landscape really is. In a way, calling the East African Rift Valley the Cradle of Humanity may be misleading because, although most of our ancestors are found within the valley floor—'the cradle'—the sides of this cradle are not gentle and protective but rather harsh jagged teeth rising thousands of metres on either side.

After 2 million years, though the rifting process slowed dramatically, there was still a lot of volcanism going on. Importantly for our story, many of the Kenyan lake basins continued to fragment due to this ongoing volcanic activity. This included the formation of the Barrier Volcano separating Lake Turkana and the Suguta Valley in the northern Kenyan Rift, which started to form about 1.4 million years ago, and finally separated the two lake basins about 0.7 million years ago (see Chapter 6 for more discussion about these lake basins). The Emuruangogolak Volcano formed about 1.3 million years ago, separating Lake Baringo and the Suguta Valley. This means that before 1.4 million years ago an interconnected lake system may have existed stretching from the Omo National Park in the north to just north of Lake Baringo in the south, a distance of over 500 km. Further south, in the Rift Valley in Tanzania, initial rifting and basin formation started about 5 million years ago. But the major phase of rift faulting occurred 1.2 million years ago, resulting in the present-day rift escarpments seen in Tanzania.

Rift systems develop from an initial triple junction, and the activity along each of the three arms varies through time. In the case of the Afar-centred triple junction, there is still a huge amount of activity in Ethiopia, and the rifting process continues in the Red Sea and the Gulf of Aden. The Somali Plate and the Nubian Plate are currently moving away from the Arabian Plate at a rate of 6–7 mm per year. However, in the lower East African Rift System the rifting process became quiescent about 200,000 years ago, and rifting has ceased. There is still tectonic activity within East Africa, but the new faults do not run north–south—they run NNW–SSE along the fractures of the very old basement rock formed during the Pan-African Orogeny, a mountain-building period over 600 million years ago.

Rifting influence on climate and vegetation

These tectonic changes are known to be associated with a variety of biotic changes, which in turn may have affected hominin evolution. Prior to the hot-spot-driven uplift, a swathe of rainforest stretched from the Congo all the way across to East Africa. This was fed by rains from the monsoon systems on either side of the African continent, bringing moisture both from the Atlantic and Indian

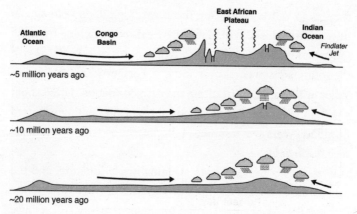

FIGURE 20 Climatic effects of the development of the East African Rift Valley system.

Oceans (Figure 20). During the Oligocene and Miocene Epochs, progressive uplift of East Africa split this pan-African rainforest. This is shown by the appearance of new 'endemic' species of plants and animals that evolved locally at about 33, 16, and 8 million years ago.

This splitting of the east and west rainforests was only the start of the huge influence rifting had on the climate of East Africa. As the uplift progressed, rain shadows were created on both the Congo and Indian Ocean side. This intensified the rainfall on the outer sides of the mountains, but led to a decrease in moisture within the Rift Valley (Figure 20). The uplift of the Tibetan Plateau and the Eastern Rift Shoulder mountain range created the Findlater Jet. This narrow, low-level, atmospheric jet stream blows diagonally across the Indian Ocean, parallel to the coasts of Somalia and Oman. The Findlater Jet accelerates the transport of moisture from the Southern Indian Ocean, past East Africa, and on to

20 million years ago

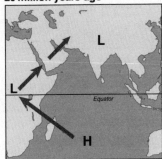

10 million years ago

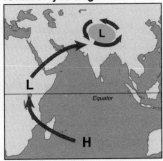

Present day

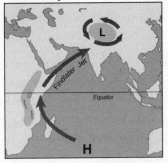

FIGURE 21 Influence of Tibetan uplift and the African Rift Valley on regional atmospheric circulation during Northern Hemisphere summer. H = high pressure area, L = low pressure area.

the Indian subcontinent, contributing to the aridification of East Africa (Figure 21).

By between 5.5 and 4 million years ago the East Africa Rift Valley had expanded to such a scale that the rain-shadow effect led to much lower rainfall, so that rainforest was no longer able to survive (Figure 22). This drying of East Africa seems to have been progressive and has continued over the past 5 million years. Evidence for this comes from pollen records and stable carbon

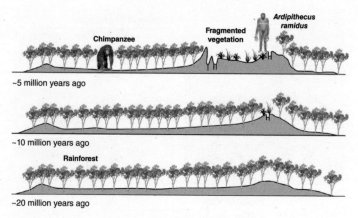

FIGURE 22 Vegetation effects of the development of the East African Rift Valley system.

isotopes measured in fossilized soil carbonates, organic matter in marine sediment, and fossilized mammal teeth. The carbon isotope records (see Box 1) show that there was a progressive shift in vegetation from mainly trees and shrubs (C_3 plants) to mainly tropical grasses (C_4 plants). This aridity trend is also supported by the results of climate model simulations. These studies demonstrate that as uplift increased, wind patterns became less east to west and more south to north, resulting in an overall decrease in regional rainfall. Hence, as elevation increased, a rain-shadow effect occurred that reduced moisture availability on the Rift Valley mountainside, producing the strong aridification trend evident in palaeoenvironmental records. Until recently this aridity was interpreted as just a long-term trend to drier conditions; but as we will see in Chapter 6, the African climate swings cyclically from wetter to drier conditions approximately every 20,000 years. We now think that what has happened over the last 5 million years is that the dry periods have got much drier and harsher, while there has been less change in the wet periods.

In addition to contributing towards the aridification of East Africa, the tectonic activity described earlier also produced numerous basins suitable for lake formation. This is shown by the increased frequency and locations of lake sediments found in East Africa over the past 10 million years. If we imagine a relatively flat East Africa 20 million years ago, then the landscape would have been dominated by large rivers, very much like the Congo Basin is today. The southward propagation of rifting, including the formation of faults and magmatic activity, is also reflected in the appearance of the earliest lake basins in the northern parts of the rift. For example, about 5 million years ago we find evidence for the beginning of lakes in Afar, Ethiopia, Omo-Turkana, and the Baringo-Bogoria Basins in northern Kenya, while the oldest lacustrine sequences in the central and southern segments of the rift in Kenya and Tanzania occur about 2.5 million years ago, at the start of the Pliocene Epoch. Palaeolakes in the northern part of the East African Rift Valley thus formed earlier than in the south. This aridification of East Africa and the shift from a riverine to lacustrine hydrology must have had a profound effect on the local environment and also on early human evolution.

Box 1: Carbon isotopes

Carbon isotopic records from carbonates and organic matter from oceans, lakes, soils, and teeth are of interest to palaeoclimatologists because they help us understand how the local to global carbon system functions. There are two stable isotopes of carbon: ^{12}C (98.89 per cent) and ^{13}C (1.11 per cent). There is no difference in the chemistry of these two isotopes—only the physical difference of weight. But the lighter isotope is used preferentially in photosynthesis, and this is an example of an isotope separation or 'fractionation' process. As a result, carbon isotopes can provide an insight into past changes in vegetation and productivity. All carbon in organic compounds in the biosphere is ultimately derived from photosynthetically produced material. Photosynthesis is a very inefficient process—despite billions of years of evolution, only about 1 per cent of sunlight falling on a leaf is converted into glucose. Hence, plants use any trick they can to make this process more efficient. One trick is to use carbon dioxide containing the lighter ^{12}C isotope, as it takes less energy to fix. Photosynthetic fixation of carbon therefore involves a large fractionation of the ^{12}C and ^{13}C isotopes. The organic matter produced by photosynthesis is very light or depleted in ^{13}C. Stable carbon isotope ratios are reported in comparison to a globally accepted standard and given a delta notation, $\delta^{13}C$, which represents the proportion of ^{13}C to ^{12}C in the sample compared to that in the standard, given as parts per thousand, or ‰. $\delta^{13}C$ values of organic carbon in land plants (higher plants) vary according to the chemical pathway of photosynthesis. Plants such as trees, shrubs, and temperate grasses use the C_3 (Calvin–Benson or non-Kranz) photosynthesis pathway and have a fractionation of between −22‰ and −30‰, compared with the isotopic composition of the atmosphere of −6‰ to −7‰. Sometime before 12 million years ago a new photosynthetic pathway evolved: the C_4 (Hatch–Slack or Kranz) pathway. This includes tropical and marsh grasses with a $\delta^{13}C$ range of

continued >

between −9‰ and −15‰ (Figure 23). C_4 plants evolved under low atmospheric carbon dioxide levels as they outcompete C_3 plants at low humidity and low CO_2 conditions—for instance, during glacial periods. This is because they have an internal 'carbon pump' system, and their primary carbon-fixing enzyme (PEP-carboxylase) does not react with oxygen—hence, increased photorespiration under low carbon dioxide conditions does not inhibit photosynthesis. This different photosynthetic pathway also helps explain why C_4 photosynthesis does not fractionate carbon dioxide as much as C_3 plants, as they do not need this additional efficiency trick. There is also a third type of plant metabolism—the CAM (Crassulacean acid metabolism)—which uses either C_3 or C_4 depending on water availability; but it is minor compared with the other two pathways.

Carbon isotopes can therefore discern whether the prevailing vegetation of a region was more C_3 (tree and shrub) or more C_4 (tropical grass dominated). These records can be generated from marine, lake, or palaeo-soil carbonates, or organic matter. Carbonate and bulk organic matter can be used directly. A more discerning approach is the use of biomarkers—particular organic compounds—which are known to originate only from terrestrial plants. For example, long-chain n-alkyl compounds are major components of epicuticular waxes from the leaves of vascular plants. These compounds are relatively resistant to degradation, which makes them suitable for use as higher plant biomarkers, and include n-alkanes, n-alkanols, n-alkanoic acids, and wax esters. Carbon isotope records can be generated from the teeth of animals, and in some cases hominins, and provide a valuable insight into the diet of early humans. For example, using $\delta^{13}C$ from fossil teeth from the Turkana Basin, Thure Cerling and his colleagues showed that *A. anamensis* was mainly forest-dwelling; *K. platyops*, *P. aethiopicus*, and early *Homo* were inhabiting grassy woodlands; while *P. boisei* was living in much more open conditions in wooded grassland.

continued>

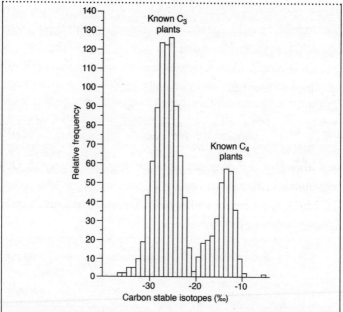

FIGURE 23 Carbon isotope differences and relative abundance from different types of plants. The difference in carbon isotope signature between C_3 plants (mainly trees and shrubs) and C_4 plants (mainly tropical grasses) allows the reconstruction of palaeo-diet and thus the dominant vegetation on the landscape.

Changing landscapes and the origins of bipedalism

There is a long history of theories of how we came down from the trees and learnt to walk upright. Starting with Jean Baptiste Lamarck in 1809 and Charles Darwin in 1871 who envisaged our ancestors learning to walk on the savannahs of Africa. This terrestrial origin theory was supported by field studies of gorillas and chimpanzees that began in the 1960s showing that their primary mode of locomotion was 'knuckle-walking' or quadruped-alism. When gorillas and chimpanzees move on the ground, they move as if on four legs by using the knuckles of their hands to support their weight. Compared to real quadrupeds this is a very inefficient way to move around. Because of the recognized genetic closeness of humans and chimpanzees, it was assumed that our common ancestor was also a knuckle-walking quadruped. So the savannah hypothesis suggested that the formation of the Rift Valley meant that dense tropical forest was slowly replaced by grassland. Forest food sources would have become fragmented, and therefore the distances between habitats increased. One evolutionary strategy to cope with the new conditions would have been to evolve the ability to walk upright, as a much more efficient way for a primate to cover large distances.

This classical view of the origins of bipedalism has recently been questioned, and we have returned to the original arboreal theory, based on a better understanding of the lifestyle and evolution of the great apes and on new hominin fossil finds. All the great apes walk bipedally, but most do so in an arboreal context. For example, orangutans spend only 2 per cent of their time using only their hind legs to move in trees, but a further 6 per cent of their time where one of other of their arms is used to support their bipedal movement. Though this is a small amount of time, it is critical to orangutans as it allows them to access smaller branches where fruit hangs. It also allows them to cross from one tree to another, avoiding the effort of climbing down to the ground, and the high predator risk involved in crossing the ground. The second line of evidence for a forest origin of bipedalism comes from the beautiful 4.4 million-year-old fossils of *Ardipithecus ramidus* found in Afar in Ethiopia. The extensive fossil remains of *A. ramidus* show that it is a bipedal hominin, but with very long arms. These long arms are clearly adapted to climbing and swinging in trees, not knuckle-walking, suggesting that *Ardipithecus ramidus* was able to travel fairly long distances using its bipedal adaptation, while retaining the ability to climb trees— essential for food gathering and escaping the large numbers of predators in East Africa.

It seems very likely therefore that our ancestors learnt to stand upright in the trees using their forelimbs for support. This also implies that the origin of bipedalism is much older than the hominin record suggests, and that it could have evolved many times. There is fossil evidence of several species of Miocene ape that show the ability to stand upright in an arboreal context. One species, *Oreopithecus bambolii*, an extinct primate found in Italy, may even have been regularly bipedal on the ground. *Oreopithecus bambolii* existed 9–7 million years ago when Tuscany and Sardinia were isolated islands, part of a chain stretching from Central Europe to North Africa. Its foot has been described as chimp-like, but its short pelvis was closer to those of hominins than those of chimpanzees and gorillas. Although it has been suggested that *Oreopithecus* was bipedal, it had a peculiar form of bipedalism, different from that of *Australopithecines*. This is because its big toe could be placed at an angle of 100° to the other toes, enabling the foot to act as a tripod in erect postures. So, although *Oreopithecus* could stand erect and walk, it would not have been able to develop particularly fast bipedal locomotion. Evidence for the hominin *Graecopithecus freybergi* has been found—a lower jaw from Greece and an upper premolar from Bulgaria. These fossil finds have been dated to about 7.2 million years ago. Unfortunately it is not possible from these finds to tell if these hominins were bipedal. But intriguingly bipedal hominin footprints have been found on the Greek island of Crete and have a tentative date of 5.7 million years ago.

The possible repeated appearance of the bipedal adaptation in different hominin lineages and its origin in a forest setting fits with our current knowledge of the landscape changes over East Africa between 10 and 5 million years ago. This is because, as the uplift and rifting of East Africa proceeded, there was no simple switch from forest to grassland: far from it. What would have happened was that the tropical forest started to fragment and a mosaic of different vegetation types appeared. Today in East Africa the vegetation ranges from cloud rainforest to arid deserts, and from open savannah to humid swampland. This fragmentation of the forest landscape would have driven adaptations within primates. One adaptation was probably for chimpanzees and gorillas to become even better at using forest resources as the total area of forest shrank. This included better tree-climbing to ensure continual access to canopy fruits while still enabling access to fall-back foods on the ground. This increased commitment to vertical climbing would have led to more flexible hips and knees, which would have favoured knuckle-walking when on the ground. The second adaptation was the likely hominin approach— adopting terrestrial bipedalism to ensure access to distant food sources but retaining the ability to climb both for food access and safety.

The legacy we have inherited from our arboreal ancestors are feet and legs that can cope with a large range of terrains, which allowed us later on to develop the ability to run very efficiently. For example, modern humans have been shown to outrun horses in different trials, such as those over 22 miles of hilly mid-Wales or 50 miles of sand dunes in the United Arab Emirates. We are not, however, fast runners compared to African predators, so the retention of long arms and powerful leg muscles to climb trees would have been valuable for our ancestors.

Over the past 10 million years, then, East Africa has changed from a rolling landscape with tropical forest covering much of the region to a mountainous fragmented landscape with almost all possible vegetation types present. In modern day Ethiopia you can travel from high mountain cloud rainforest to lowland arid desert in the space of less than 50 miles. The development of the Rift Valley between 10 and 5 million years ago created two mountain ranges, the Rift shoulders, up to 4 km high in places, running roughly north to south. These mountains created rain shadows and deflected atmospheric moisture westward back into the Congo and eastward across to the Indian subcontinent. During this period the region experienced a huge environmental change. As we will see in Chapter 5, this change was compounded by the re-evolution of tropical grasses, adapted to very dry conditions, which in part drove the emergence of the savannah. The East African primates responded to these massive environmental changes in two ways: chimpanzees and gorillas became better at tree-climbing, while the hominins enhanced their arboreal bipedalism to enable them to move around efficiently between the fragmented food sources on the ground.

5

Global Climate Change

Massive changes have occurred in East Africa over the past 10 million years due to tectonic uplift and rifting. At the same time revolutionary changes were occurring in the Earth's climate system as it switched from a 'greenhouse' to an 'icehouse' world. Fifty million years ago the Earth was a very different place. The world was both warmer and wetter, with rainforest extending all the way up to Canada and all the way down to Patagonia. In this chapter we examine how it became transformed from this lush vibrant state to the ice-locked cool planet we have today. We will explore what triggered the series of ice ages and the effects these changes had on early human evolution in Africa.

Super-lush Earth

Before we investigate what caused the massive cooling of the Earth, we should consider one key climatic event which occurred during the period when Earth was hot, warm, and lush, that had a profound effect on mammal evolution. The Palaeocene–Eocene

Thermal Maximum (PETM) is a climatic event that defines the temporal boundary between the Palaeocene and Eocene Epochs in the rock record (Figure 24). The event occurred about 55.8 million years ago and lasted about 170,000 years. During this hothouse event, scientists think that 1,500 gigatonnes of gas hydrates may have been released. Gas hydrates, or clathrates, are mixtures of water and methane which are solid at low temperatures and high pressures. They are made up of cages of water molecules which hold individual molecules of methane or other gases. The methane comes from decaying organic matter found deep in ocean sediments and in soils beneath permafrost. Gas hydrate reservoirs can be unstable, as an increase in temperature or decrease in pressure can cause them to destabilize, releasing the trapped methane. It is thought that the warmth of the Palaeocene period caused the breakdown of the global gas hydrate reserves, releasing the trapped methane. Methane in the atmosphere is twenty-one times more powerful as a greenhouse gas than carbon dioxide, and this catastrophic release may be the cause of the observed 5°C temperature rise in global temperatures. On land, the temperature rises were even larger.

During the PETM the atmosphere became more humid, and mangroves and rainforests spread as far north as England and Belgium and as far south as Tasmania and New Zealand. Turtles, hippopotami, alligators, and palm trees have been found at Ellesmere Island in the Canadian Arctic. The Arctic Ocean was very warm but stagnant, with sea temperatures rising from 18°C to over 23°C. This was a period of rapid extinctions and evolutionary change as local species struggled to adapt to changing environments while facing fierce competition from migrating species. It is during the PETM that we have the first fossil evidence

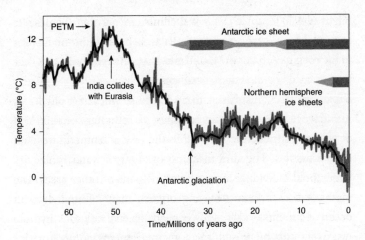

FIGURE 24 Global temperatures through the Cenozoic.

for true primates (*Teilhardina brandti* and *Teilhardina belgica*) and the first social monkeys. It seems that one adaptation to the rapid climate changes during the PETM was for primates to band together into large social groups—an essential development, as we think sociality is a key driver in hominin brain expansion. In addition, the ancestors of many modern mammals such as hoofed animals (pygmy horses and elk), tapirs, rodents, bats, owls, elephants, and early whales also appeared during or shortly after the PETM, heralding a new period of mammal diversification—the Eocene Epoch.

Growth of ice sheets

If you compare a map of the world about 50 or 60 million years ago to today, they seem identical—until you look in detail

(Figure 25). In Chapter 3 we saw that movements of the continents around the face of the planet are very slow, but that minor changes in location have had a profound effect on global climate. Over the past 60 million years these small changes have moved the Earth's climate from a greenhouse to an icehouse world. Of particular importance is having continents on or surrounding the poles and a means of cooling them down. In the case of Antarctica the ice did not start building up until about 35 million years ago (Figure 24). Prior to that Antarctica was covered in lush temperate forest; even bones of dinosaurs have been found there before they went extinct 65 million years ago. What changed 35 million years ago was a culmination of minor tectonic movements. South America and Australia were slowly moving away from Antarctica, as the large southern continent of Gondwana fragmented. About 35 million years ago the ocean opened up between Australia and Antarctica. This was followed about 30 million years ago by the opening of the Drake Passage between South America and Antarctica, one of the most feared stretches of ocean. The opening up of these seemingly small ocean gateways between the continents produced an ocean, the Southern Ocean, which could circulate completely around Antarctica, continually drawing heat from the continent and releasing it into the Atlantic, Indian, and Pacific Oceans with which it mixes. Such has been the efficiency of this process that there is enough ice on Antarctica today to produce a global sea-level rise of over 65 metres if it were all to melt—enough to cover the head of the Statute of Liberty. This tectonic cause of the glaciation of Antarctica is also the reason that scientists are confident that the Eastern Antarctic ice sheet, with the potential of producing about 60 metres of sea-level rise, will not melt due to future climate change.

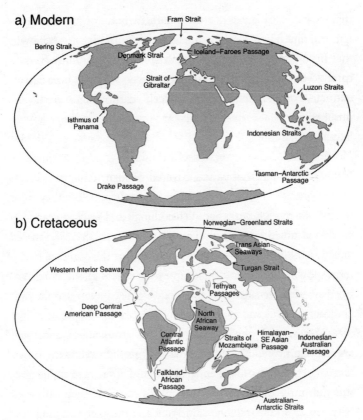

FIGURE 25 Ocean gateways both today and during the Cretaceous.

The ice-locked Antarctica of 30 million years ago did not, however, last long. Between 25 and 10 million years ago Antarctica ceased to be completely covered with ice. This raises some questions. Why did the world start to cool all over again 10 million years ago? And why did the ice start building up in the northern hemisphere 2.5 million years ago? Palaeoclimatologists believe that relatively low levels of atmospheric carbon dioxide are essential to maintaining a cold planet. Computer models have shown that if you have high levels of atmospheric carbon dioxide you cannot get ice to form on Antarctica, even with the ocean heat extractor. So somehow atmospheric carbon dioxide levels dropped and ice started building in Antarctica, and then much later in the Arctic.

What caused the big freeze?

In 1988, palaeoceanographer Bill Ruddiman and his then graduate student Maureen Raymo, at the Lamont-Doherty Earth Observatory, wrote an influential paper. They suggested that the observed long-term global cooling and build-up of ice sheets in the northern hemisphere were caused by the uplift of the Tibetan–Himalayan and Sierran–Coloradan regions. As we saw in Chapter 3, huge plateaux can alter the circulation of the atmosphere, and they argued that this cooled the northern hemisphere, allowing snow and ice to build up. However, what they did not realize at the time was that most of the Himalayan uplift occurred much earlier— between 20 and 17 million years ago—and thus too early to have been the direct cause of the ice in the north. But Maureen Raymo then came up with the startling suggestion that this uplift may have caused massively increased erosion, using up atmospheric carbon dioxide in the process. When you make a mountain range you also produce a rain shadow: one side of the mountain has a lot more rain on it as air is forced up and over the mountain. This is also why mountains erode much faster than gently rolling hills. She argued that this extra rainwater and CO_2 from the atmosphere formed a weak carbonic acid solution, which dissolves rocks. But interestingly, only the weathering of silicate minerals makes a difference to atmospheric CO_2 levels, as weathering of carbonate rocks by carbonic acid returns CO_2 to the atmosphere. As much of the Himalayas is made up of silicate rocks, their chemical weathering could remove a lot of atmospheric carbon dioxide. The resultant minerals dissolved in the rainwater are then washed into the oceans and used by marine plankton to make shells out of the calcium carbonate. The calcite skeletal remains of the marine biota are ultimately deposited as deep-sea sediments and hence lost from the global carbon cycle for the duration of the life cycle of the oceanic crust on which they were deposited. It's a fast-track way of getting atmospheric carbon dioxide out of the atmosphere and dumping it at the bottom of the ocean. Geological evidence for long-term changes in atmospheric carbon dioxide does support the theory that it has dropped significantly over the past 20 million years. The only problem scientists have with this theory is what stops the process. With the amount of rock in the Tibetan Plateau that has been eroded over the past 20 million years, all the carbon dioxide in the atmosphere should have been stripped out. So there must be other natural mechanisms which help to maintain the balance of carbon dioxide in the atmosphere, as the long-term concentration of carbon dioxide is a balance between what is removed by weathering and deposition in the deep ocean and what is added when it is recycled at subduction zones and emitted by volcanoes.

With atmospheric carbon dioxide lowering between 10 and 5 million years ago the Greenland ice sheet started to build up. So, by 5 million years ago, we had huge ice sheets on both Antarctica and Greenland, very much like today. The great ice ages, when huge ice sheets waxed and waned on North America, Northern

Europe, New Zealand, and Patagonia, did not start until 2.5 million years ago.

The Great Salt Crisis

Between 11 and 7 million years ago, tectonic changes caused the eastern end of the Tethys Sea gradually to close. This led to the shifting of the African summer monsoons, and we have evidence for the first sand dunes in North Africa about 7 million years ago. This is also when we have evidence for possibly the earliest hominin, *Sahelanthropus tchadensis*, in Chad. About 6 million years ago, the gradual tectonic changes in western parts of the Mediterranean Sea resulted in the closure of the precursor of the Strait of Gibraltar. This led to the transient isolation of the Mediterranean Sea from the Atlantic Ocean. At 5.6 million years ago, the strait closed for the last time and, because of the generally dry climate conditions, within a millennium the Mediterranean basin had completely dried out, leaving a deep dry basin in some places between 3 to 5 km below the world ocean level (Figure 26). There were a few hypersaline Dead Sea-like lakes. Around 5.5 million years ago, less dry climatic conditions allowed the basin to resume receiving more fresh water from rivers, and brackish water lakes—like the Caspian Sea today—were created, becoming progressively less hypersaline, until the Strait of Gibraltar finally reopened at 5.33 million years ago. During the isolation of the Mediterranean region, desiccation led to the deposition of vast salt deposits. Imagine a huge version of the Dead Sea where a few metres of seawater covers a vast area. This event, called the Messinian Salinity

Crisis, was a global climate event because nearly 6 per cent of all dissolved salts in the world's oceans were removed. By five and a half million years ago, the Mediterranean Sea was completely isolated and a salt desert (Figure 26). This was roughly at the same time as palaeoclimate records indicate that the northern hemisphere was starting to glaciate. This desiccation of the Mediterranean Sea would have had a profound effect on the climate of North Africa. Intriguingly, the first evidence for *Ardipithecus kadabba* is also dated at 5.6 million years ago. It is tempting to suggest that the drying out of the Mediterranean region, and thus East Africa, led to the first known truly bipedal hominin. But we have so few fossils from 10 and 4.4 million

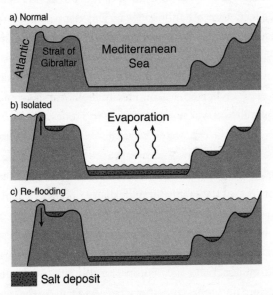

FIGURE 26 Drying out, salt deposit formation, and reflooding of the Mediterranean Sea approximately 5 million years ago.

years ago that this would be pure speculation; fun, but not very scientific.

At about 5.33 million years ago, the Strait of Gibraltar opened up and caused the Terminal Messinian Flood, which is also called The Zanclean Flood or Zanclean Deluge. Scientists have envisaged an immense waterfall higher than today's Angel Falls in Venezuela (979 m), and far more powerful than either the Iguazu Falls on the boundary between Argentina and Brazil or the Niagara Falls on the boundary between Canada and the USA. More recent studies of the underground structures at the Gibraltar Strait show that the flooding channel may have descended in a more gradual way to the dry Mediterranean. The flood could have occurred over months or a couple of years, but meant lots and lots of dissolved salt was pumped back into the world's oceans via the Mediterranean–Atlantic gateway. This stopped the start of the great ice ages at this time—all because of how oceans circulate. The Gulf Stream in the Atlantic Ocean not only keeps Europe warm, but also drives the deep ocean circulation and keeps the whole planet relatively warm. Five million years ago this deep ocean circulation was not as strong as it is today. This is because fresher Pacific Ocean water was still able to leak through the Panama ocean gateway, which is discussed later in this chapter. The sudden massive increase in salt due to the Terminal Messinian Flood increased the salinity and the density of the North Atlantic Ocean, ensuring a very vigorous Gulf Stream and sinking water in the Nordic seas. With all this tropical heat being efficiently pumped northwards, the slide into the great ice ages was halted about 5 million years ago. We had to wait another 2.5 million years before the global climate system was again on the threshold of a great ice age.

Appearance of tropical grasses

The evolution, emergence, and expansion of grasses using the C_4 photosynthetic pathway (Box 1), which took place during the Mid to Late Miocene, is thought to have been driven by a lowering of atmospheric carbon dioxide levels. This was a global climate event, as the evolution of tropical C_4 grass-dominated biomes had a long-lasting impact on continental biota, including major shifts in vegetation structure and, in Africa, accelerated forest shrinkage and the emergence of more open landscapes accompanied by large-scale evolutionary shifts in animal communities. Although the theoretical threshold in atmospheric carbon dioxide concentrations for the evolution of C_4 photosynthetic pathway was breached ~30 million years ago, evidence for its existence starts 12 million years ago in Asia. Carbon isotope evidence from fossil soils and biogenic carbonate suggest that tropical grass-dominated biomes emerged between 8 and 7 million years ago in Africa.

The Panama Paradox

Another important tectonic control, which geologists believe to be a trigger for the great ice ages, is the closure of the Pacific–Caribbean gateway. The Panama ocean gateway began to close 4.5 million years ago, and finally closed around 2 million years ago. However, its closure creates a paradox, as it would have both helped and hindered the start of the great ice ages. First, the reduced inflow of Pacific surface water to the Caribbean would have increased the salinity of the Caribbean because Pacific Ocean water is fresher than its counterpart in the North Atlantic Ocean. This would have increased the salinity of water carried northward by the Gulf Stream and North Atlantic Current and, as we noted earlier, this would have enhanced deep-water formation. The increased strength of the Gulf Stream and deep-water formation would have worked against the start of the great ice age as it enhances the oceanic heat transport to the high latitudes and would have opposed ice sheet formation. So, after the aborted attempt to start the great ice age about 5 million years ago, the progressive closure of the Panama ocean gateway kept increasing the heat transported northward, keeping the chill at bay. But here is the paradox, because two things are needed to build large ice sheets: cold temperatures and lots of moisture. The enhanced Gulf Stream also pumped a lot more moisture northward, ready to stimulate the formation of ice sheets. And the building of large ice sheets in the northern hemisphere could start at a warmer temperature because of all the extra moisture being pumped northward, ready to fall as snow and to build up ice sheets.

Why 2.5 million years ago?

Tectonic forcing alone cannot explain the amazingly fast intensification of northern hemisphere glaciation (Figure 27). My own research using ocean sediments suggests that there were three main steps in the transition to the great ice ages. The evidence is based on when rock fragments ripped off the continent by ice were deposited in the adjacent ocean basin by icebergs. First, ice sheets started growing in the Eurasian Arctic and North East Asia regions approximately 2.74 million years ago, with some evidence of growth of the North East American ice sheet. Second, an ice sheet started to build up on Alaska 2.7 million years ago. And third, the biggest ice sheet of them all on the North East American continent reached its maximum size 2.54 million years ago. So, in less than 200,000 years, we go from the warm balmy conditions of the early Pliocene to the great ice ages.

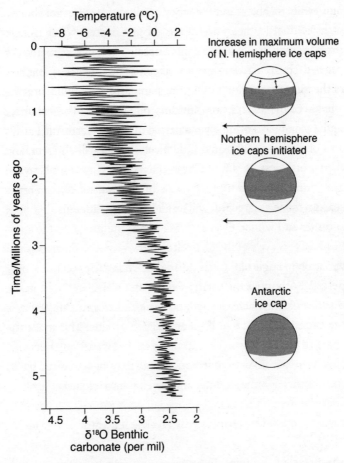

FIGURE 27 Oxygen isotopes showing the step-like changes in global ice volume over the past 5 million years.

The timing of the start of the intensification of northern hemisphere glaciation must have had another cause. It has been suggested that changes in orbital forcing—i.e. changes in the way the Earth spins round the Sun—may have been an important mechanism contributing to the global cooling. The details of the Earth's numerous wobbles, and how they cause the waxing and waning of individual ice ages, are discussed in Chapter 6. But though these individual wobbles are on the scale of tens of thousands of years, there are much longer variations as well. For example, one of the most important is obliquity or angle of the tilt, which is the wobble of the Earth's axis of rotation up and down with respect to the plane of its orbit. Over a period of 41,000 years, the Earth's axis of rotation will lean a little bit more towards the Sun, and then a little less. It is not a large change, varying from between 21.8° and 24.4°. The tilt of the axis of rotation produces Earth's seasons, hence a larger tilt will result in greater difference between summer and winter. Over a period of 1.25 million years the amplitude of this cycle of the tilt itself changes—that is, the maximum variation in angle of the axis during the cycle varies on this longer period. Both times the Earth tried to enter a glaciated state in the northern hemisphere, at 5 million years and 2.5 millions year ago, the variation of tilt had been at its largest value. This made the changes in each season exceptionally large. Of particular importance were the cold summers in the north, which allowed ice to survive through the summer and develop into ice sheets.

The tropics react to the ice ages

The onset of the intensification of northern hemisphere glaciation between 2.75 and 2.55 million years ago has, for the past thirty-five years, been cited as the main cause for the appearance of *Homo* in East Africa. With the recent discovery of a possible species of *Homo* at 2.8 million years, and *Homo habilis* appearing at 2.35 million years, the two events do seem to overlap. However, there is little evidence for major environmental change in Africa at this time. It has also been argued that *Homo habilis* was not that different to those members of the *Australopithecus* genus. With the discovery of stone tools at 3.3 million years the special place of *Homo habilis* has gone. This should help discount the obsession with the view that intensification of northern hemisphere glaciation was the great cause of human evolution. It would appear that it was half a million years after the start of the great ice ages that things started to change in the tropics.

Before 1.9–1.8 million years ago it seems there was very little east–west sea surface temperature gradient in the Pacific Ocean, but there was afterwards. This indicates a switch in the tropics and subtropics to a modern mode of circulation, with relatively strong Walker circulation and cool subtropical temperatures. The Walker circulation is the atmospheric east–west component of the Hadley cell and is instrumental in controlling rainfall in the topics. It is also a key element in the El Niño-Southern Oscillation or ENSO (see Box 2). So before 2 million years ago, ENSO may not have existed in its modern form because there was a relatively weak Walker circulation. The development of the Walker circulation after 1.9 million years ago also seems to have been linked to major changes in the African environment. Carbon isotopes from fossil mammals suggest that, although there was a general trend towards more open environments after 3 million years ago, the most significant environmental change to open, grassy landscapes occurred after 2 million years ago, rather than 2.6–2.4

million years ago, as earlier research had suggested. The reanaly-sis of terrestrial dust records found in marine sediment from the Arabian Sea, the eastern Mediterranean Sea, and off subtropical West Africa all suggest an increase in aridity and great climatic variability on the continent after 1.9–1.5 million years ago. As we will see in Chapter 7, there is also evidence for large, deep, but fluctuating lakes occurring throughout East Africa. And from 1.8 million years ago H. erectus, with an 80 per cent increase in brain size from previous hominins, evolved in East Africa, and the first hominin migrations out of Africa began.

Box 2: El Niño-Southern Oscillation

One of the most important and mysterious elements in global climate is the periodic switching of direction and intensity of ocean currents and winds in the Pacific region. Originally known as El Niño ('Christ child' in Spanish) because it usually appears at Christmas, it is now generally known as ENSO (El Niño-Southern Oscillation), and it typically occurs every three to seven years. It may last from several months to more than a year. ENSO is an oscillation between three climate states: the 'normal' conditions, La Niña, and 'El Niño'. ENSO has been linked to changes in monsoons, storm patterns, and the occurrence of droughts throughout the world. For example, the prolonged ENSO event in 1997–8 caused severe weather events all over the Earth, including droughts in East Africa, northern India, north-east Brazil, Australia, Indonesia, and Southern US, and heavy rains in California, parts of South America, the Pacific, Sri Lanka, and east-central Africa.

In an El Niño event, the warm surface water in the Western Pacific moves eastward to the centre of the Pacific Ocean (Figure 28). Hence, the strong convection cell or column of warm rising air is much closer

continued >

to South America. Consequently the trade winds are much weaker and the ocean currents crossing the Pacific Ocean are weakened. This reduces the amount of cold nutrient-rich water upwelling off the coast of South America, and without those nutrients the abundance of life in the ocean declines and fish catches are dramatically reduced. This massive shift in ocean currents and the position of the rising warm air changes the direction of the jet streams and upsets the weather in North

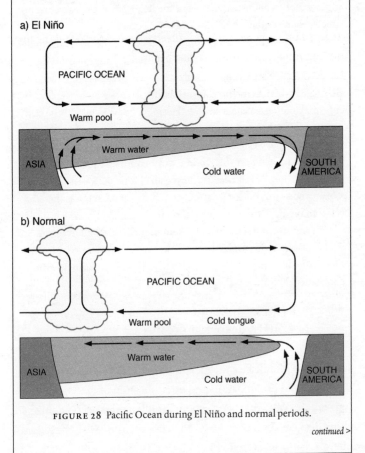

FIGURE 28 Pacific Ocean during El Niño and normal periods.

continued >

95

America, Africa, and the rest of the world. However, if you ask what causes El Niño the answer is a sort of chicken-and-egg one. Does the westward ocean current across the Pacific reduce in strength, allowing the warm pool to spread eastward moving with the wind system? Or does the wind system relax in strength, reducing the ocean currents, which then allow the warm pool to move eastwards? Many scientists believe that long-period waves in the Pacific Ocean that move between South America and Australia over time help shift the ocean currents which produce either an El Niño or La Niña event.

A La Niña event is a more extreme version of the 'normal' conditions. Under normal conditions, the Pacific warm pool is in the Western Pacific and there are strong westerly winds and ocean currents keeping it there. This results in upwelling off South America and lots of nutrients, making for excellent fishing. During a La Niña period the temperature difference between the Western and Eastern Pacific becomes extreme, and the westerly winds and ocean currents are enhanced. The impacts of La Niña on the world's weather are less predictable than those of El Niño. This is because during an El Niño period the Pacific jet stream and storm tracks get stronger and straighter and it is therefore easier to predict its effects. La Niña, on the other hand, weakens the jet stream and storm tracks, making them more looped and irregular. So the behaviour of the atmosphere, and in particular of storms, becomes more difficult to predict. In general, where El Niño is warm, La Niña is cool; where El Niño is wet, La Niña is dry.

Longer, deeper, more intense ice ages

The Early to Middle Pleistocene transition (EMPT) is the name given to the marked prolongation and intensification of glacial–interglacial climate cycles that began about 1 million years ago and finished about 800,000 years ago (Figure 27). Prior to the EMPT, the glacial–interglacial cycles seem to have occurred every 41,000 years, which corresponds to the slow changes in the tilt of the Earth. After about 800,000 years ago the glacial–interglacial cycles seem to have become much longer, averaging over 100,000 years. The shape of these cycles also changes. Before the EMPT the transition between glacial and interglacial periods was smooth and the world seems to have spent a roughly equal amount of time in each climate. After the EMPT the cycles become saw-toothed, with ice building up over a period of 80,000 years to produce deep intense ice ages, followed by rapid deglaciation or loss of all that ice within 4,000 years. The climate then stays in an interglacial period like our current climate for about 10,000 years before descending back into an ice age. One suggestion for this saw-toothed pattern is that the much larger ice sheets are very unstable and therefore with a slight change in climate they collapse rapidly and the whole climate system rebounds back into an interglacial period. It is also interesting to note that prior to the EMPT the ice age cycles appear to have had very little effect on the tropics, which seemed to have a completely different climate cycle lasting only 21,000 years, driven by different parts of the orbital forcing system. After the EMPT the glacial–interglacial cycles seem to have had an increasing influence on the tropics, and in particular Africa. I am sure you will also have noticed that the end of the EMPT, when the much longer glacial–interglacial cycles were established, occurred around the time when we find the first evidence for *H. heidelbergensis*.

Unstable ice ages

In many ways, ice ages or glacial periods should really be called 'climate rollercoasters', because ice sheets are naturally unstable and cause the climate to veer violently from one state to another as the ice sheets dramatically collapse and then regrow. Most of the variations occur on the millennial timescale, but the start of these extreme events can occur in as little as three years. The most impressive of these are the Heinrich events, named by Professor Wally Broecker, a palaeoceanographer at Lamont-Doherty Earth Observatory, after a paper by Hartmut Heinrich describing them in 1988. Heinrich events are massive collapses of the North American Laurentide ice sheet that resulted in millions of tonnes of ice being poured into the North Atlantic Ocean. Wally Broecker, in his usual flamboyant way, described them as armadas of icebergs floating from North America across the Atlantic Ocean to Europe. Huge gouges have been found on the north French coast where these enormous icebergs ran aground. These Heinrich events occurred against the general background of unstable glacial climate, and represent the brief expression of the most extreme glacial conditions around the North Atlantic region. The Heinrich events are evident in Greenland ice core records as a further 2–3°C drop in temperature from the already cold glacial climate. The Heinrich events have been found to have had a global impact, with evidence for major climate changes described from as far afield as the Americas, China, the Arabian Sea, and East Africa. During these events around the North Atlantic region much colder conditions are found both in North America and Europe. In the North Atlantic Ocean the huge number of melting icebergs added so much cold fresh water that sea surface temperatures and salinity were reduced to the extent that the surface water could not sink. This stopped all deep-water formation in the North Atlantic Ocean, switching off the global ocean conveyor belt. Heinrich events seem to occur every 13,000 to 7,000 years during an ice age. We now know that between these massive Heinrich events there are smaller events occurring at about every 1,500 years, which are referred to as Dansgaard–Oescheger events or cycles. There is evidence from African lake sediments that these millennial climate cycles may have been influencing African climate as far back as 2.7 million years, further complicating our picture of the climate in which our ancestors lived.

During the period of early human evolution in Africa, then, there were five major transitions or climate events that had a significant influence on African climate. The first was the emergence and expansion of C_4 grass-dominated biomes from 8 million years ago onwards, which would have helped dry out the region and led to the creation of the iconic savannahs of Africa. Second was the Messinian Salinity Crisis between 5.95 and 5.33 million years ago, and the desiccation of the whole Mediterranean region. The third transition was the intensification of northern hemisphere glaciation and the start of the great ice ages between 2.75 and 2.55 million years ago. The fourth was the development of the Walker circulation between 1.9 and 1.7 million years ago and the development of a modern ENSO system, both of which seem to have caused significant changes in the climate of East Africa. And the fifth was the Early–Middle Pleistocene transition, between 1 and 0.8 million years ago, when Africa first became influenced by the northern hemisphere-driven glacial–interglacial cycles and the millennial-scale climate cycles associated with the collapse of giant North American ice sheets. Many of the climate events coincide with major changes in human evolution, but how and why these changes occurred will be discussed in later chapters.

In Chapter 6 we will consider how changes to the Earth's orbit have influenced climate—in particular, the climate of Africa.

6

Celestial Mechanics

—

Tectonics is the main driving force for long-term global and regional changes in climate. In the shorter term there is another major control—celestial mechanics. The Earth's orbit around the Sun and the angle of its axis of rotation vary or 'wobble'. We now know that this 'orbital forcing' drove the glacial–interglacial cycles that are fundamental characteristics of the Quaternary Period, the past 2.5 million years. More recently we have found that orbital forcing also has a profound effect on the seasons in the tropics, changing the occurrence of rainfall, and in East Africa has led to the expansion and shrinking of large, deep-water lakes within the Rift Valley.

Orbital forcing

Orbital forcing is one of the hardest subjects I teach, and despite my repeated efforts we have never been able to show it properly on TV. So please stick with me through the next section and I will take you though it step by step. There are three main orbital parameters or wobbles: eccentricity, obliquity (tilt), and precession, which have a significant effect on the long-term climate of the Earth.

Eccentricity describes the shape of the Earth's orbit around the Sun. It varies from nearly a circle to an ellipse, due to the gravitational force exerted by Jupiter. The Earth's orbit varies from nearly circular, with an eccentricity of 0.005, to quite elliptical, with an eccentricity of 0.06 (Figure 29). Currently we have an eccentricity of 0.0174—nearly its minimum. This variation occurs over a period of about 96,000 years with an additional long cycle of about 413,000 years (Figure 30). Described another way, the long axis of the ellipse varies in length over time. Imagine the Earth's orbit around the Sun as a perfectly circular rubber band. You put two fingers into the rubber band and slowly expand it so it makes an ellipse shape. When you release your fingers the rubber band goes back to a circle. Today, the Earth's orbit is a small ellipse and the Earth is closest to the Sun on 3 January, at about 146 million km; this position is known as the perihelion. On 4 July the Earth is most distant from the Sun—about 156 million km at the aphelion. Changes in eccentricity cause only minor variations in total insolation (solar energy that falls on the Earth's surface), but can have a significant effect on the insolation received throughout the year and thus a big seasonal effect. If the orbit of the Earth were perfectly circular there would be no seasonal variation in solar insolation. Today, the average amount of radiation received by the Earth at perihelion is ~351 Watts per square metre (W/m^2), and 329 W/m^2 at aphelion. This represents a difference of ~6 per cent, but at times of maximum eccentricity (ellipse length) over the past 5 million years the difference could have been as large as 30 per cent.

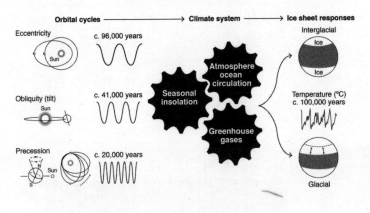

FIGURE 29 Orbital eccentricity, tilt, and precession.

Obliquity, or the tilt of the Earth's axis of rotation with respect to the plane of its orbit, varies between 21.8° and 24.4° over a period of 41,000 years (Figure 29). It is the tilt of the axis of rotation that causes the seasons. Currently the Earth is tilted at 23.44° from its orbital plane, roughly halfway between its extreme values. The larger the obliquity, the greater the difference between the insolation received in summer and winter. As we saw earlier, before about 1 million years ago, global climate was dominated by this 41,000-year cycle. Even before the great ice ages 2.5 million years ago the climate swung gently between warmer and cooler conditions, driven by obliquity. Obliquity also has a long-term cycle, with the variations in tilt slowly increasing and then decreasing over 1.5 million years (Figure 30).

Precession is the most complicated forcing of the three main orbital parameters, but has the most effect on the tropics. Precession has one component relating to the elliptical orbit of the Earth (eccentricity) and its axis of rotation (Figure 29). The

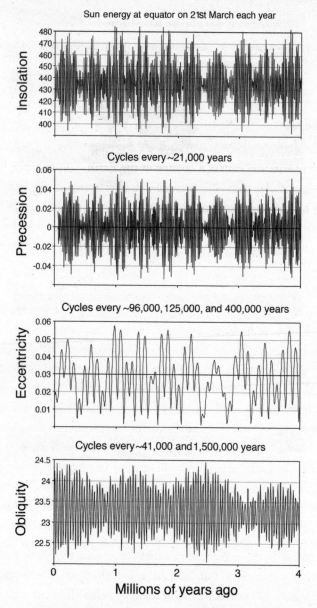

FIGURE 30 Long-term changes in the orbital parameters.

Earth's rotational axis moves around a full circle, or precesses, every 26,000 years. This gyroscopic motion is due to the tidal forces exerted by the Sun and the Moon on the solid Earth, which are amplified as the Earth is shaped as an oblate spheroid rather than a sphere. This is similar to the gyrations of the rotational axis or 'plunger' of a toy spinning-top. Precession of the axis of rotation causes a change in the Earth–Sun distance for any particular date—for example, the beginning of the northern hemisphere summer. However, this will only influence climate if the orbit of the Earth is an ellipse, because when the Earth's orbit around the Sun is a perfect circle there would be no difference in the Earth–Sun distance. So the effect of precession of the axis of rotation is modulated by eccentricity, and combining these two orbital parameters results in a periodicity in the effect of 19,000 years.

The second component relates to the precession of the whole of the Earth's orbit around the Sun (Figure 29). Every 105,000 years the Earth's whole orbital rotation around the Sun swings around the Sun. Imagine the Earth's orbital path around the Sun as hula hoop, and that you are one of those gifted people who can spin a hoop around on one ankle without falling down. That spin of the hula hoop is similar to what happens to the Earth's orbit around the Sun. This effect also modulates the influence of the precession of the axis of rotation, creating a periodicity of 23,000 years. These two periodicities combine so that perihelion coincides with the summer season in each hemisphere—on average every 21,700 years. Precession has the most significant impact in the tropics, because obliquity has no direct impact at the equator. So although obliquity clearly influences high-latitude climate change, which may ultimately influence the tropics, the direct effects of insolation in the tropics are due to eccentricity-modulated precession alone. By modelling the positions of the planets in the Solar System, astronomers can combine the effects of eccentricity, obliquity, and precession to provide the means of calculating insolation for any latitude back through time. Of course, small variations in the different gravitational influences disrupt the calculations and so they become less accurate the further back we try to calculate insolation on Earth.

Waxing and waning of the great ice ages

We know about the orbital parameters because of nearly five decades of scientific endeavour to understand the driving forces behind the great glacial–interglacial cycles. The waxing and waning of the huge continental ice sheets were initiated by changes in the Earth's orbit around the Sun. In 1949, Milutin Milankovitch, a brilliant Serbian mathematician and climatologist, was the first scientist to combine the effects of all three orbital parameters to calculate the solar energy received by the Earth. He suggested that summer insolation at 65°N, which is just south of the Arctic Circle, was critical in controlling glacial–interglacial cycles (Figure 31). He argued that, if the summer insolation was reduced enough, then ice could survive through the summer, start to build up, and eventually produce an ice sheet.

Orbital forcing does have a large influence on this summer insolation. The maximum change in solar radiation in the past 0.5 million years is equivalent to reducing the amount of summer radiation received today at 65°N to that received now over 550 km to the north at 77°N. In simplistic terms, it brings the current ice limit in mid-Norway down to the latitude of

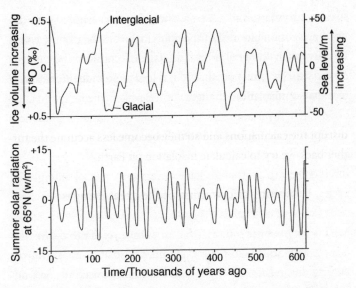

FIGURE 31 Insolation at 65°N compared with global sea level or global ice volume.

mid-Scotland. These lows in 65°N insolation are caused by eccentricity elongating the summer Earth–Sun distance, obliquity being shallow and precession placing the summer season at the longest Earth–Sun distance produced by eccentricity. The reason why it is 65°N and not 65°S, which controlled global climate in the Cenozoic, is because of the current configuration of the continents: thus any ice that builds up in the northern hemisphere has lots of continents to grow upon. In contrast, in the southern hemisphere, the ice growth is limited by the Southern Ocean, because any extra ice produced on Antarctica falls into the ocean and is swept away to warmer seas. It is almost as though Antarctica is already full up with ice and unable to accept any more. So the conventional view of glaciation is that low summer solar energy in the temperate northern hemisphere allows ice to survive summers, and thus ice sheets start to build up on the northern continents. But this simple clockwork view of the world is really much more complicated, as the effects of orbital changes on the seasons is very small, and it is feedbacks in the climate system that amplify these changes.

Glacial–interglacial cycles

It is incorrect to say that orbital forcing causes glacial–interglacial cycles. First, there isn't a one-to-one relationship between cause and effect. For example, the position of the Earth's orbital parameters is very similar today to that of 21,000 years ago during the last ice age. So it is not the exact orbital position that controls climate, but rather the changes in the orbital positions. Second, orbital forcing in itself is insufficient to drive the observed glacial–interglacial variability in climate. Instead, the Earth system amplifies and transforms the changes in solar energy received at the Earth's surface through various feedback mechanisms. For example, let us start with building an ice age. The first thing to happen is a slight reduction in summer temperatures. As snow and ice accumulate due to initial changes in summer temperature, the albedo—the reflection of sunlight back into space—increases. Reflecting more sunlight back into space suppresses local temperatures; this promotes the accumulation of more snow and ice increasing the albedo of the region, producing the so-called 'ice–albedo' feedback. So once you have a small ice sheet, it changes the environment around it to make more snow and ice, and will get bigger and bigger.

Another feedback cycle is triggered when the ice sheets, particularly the Laurentide ice sheet on North America, become big enough to deflect the atmospheric planetary waves (see Figure 14b). This changes the storm path across the North Atlantic Ocean and prevents the Gulf Stream and North Atlantic Drift penetrating as far north as it does today. This surface ocean change, combined with the general increase in melt-water in the Nordic Seas and Atlantic Ocean due to the presence of large continental ice sheets, ultimately leads to reduction in the production of deep water. Deep-water production in the Greenland and Labrador Seas is the heartbeat of the modern climate. By reducing the formation of deep water it reduces the amount of warm water pulled northwards, all of which leads to increased cooling in the northern hemisphere and expansion of the ice sheets.

There is currently debate among palaeoclimatologists about the relative roles of the 'physical climate' feedbacks described here and of greenhouse gases in the atmosphere. Air bubbles trapped in polar ice have shown us that carbon dioxide dropped by a third, and methane by a half, during each glacial period. These changes would have compounded the cooling that occurred during each glacial period, helping to build more ice. The argument continues: do changes in the Earth's orbit affect the production of greenhouse gases, cooling down the Earth to make the northern hemisphere continents susceptible to the build-up of large ice sheets? Or do changes in the Earth's orbit start to build up large ice sheets in the northern hemisphere that then change global climate and reduce the production of greenhouse gases, prolonging and deepening the ice age? The jury is still out on this one. However, what we do know is that greenhouse gases played a critical role in glacial–interglacial cycles. We also know that changes in greenhouse gas concentration always comes before changes in global temperatures.

An important question is why don't these feedbacks end up becoming runaway and freezing the whole Earth? They are prevented from doing so by 'moisture limitation'. To build an ice sheet you need the climate to be cold and wet. As the warm surface water is forced further and further south, the supply of the moisture required to build ice sheets decreases. So the ice sheets, by changing the atmospheric and ocean circulation, end up starving themselves of moisture—a negative feedback loop that limits the effects of the positive feedbacks.

In the past million years it took up to 80,000 years to build up ice sheets to reach the maximum extent of ice. The last time this occurred was about 21,000 years ago. However, getting rid of the ice is much quicker. This is called deglaciation and it usually took a maximum of only 4,000 years. Now this deglaciation is triggered by an increase in solar energy received in the summer at about 65°N, encouraging the northern hemisphere ice sheets to melt slightly. The rise of atmospheric carbon dioxide and methane promotes warming globally and encourages the melting of the large continental ice sheets. But these processes have to work against the ice sheets' albedo effect, which produces a microclimate that tries to keep them intact. What causes the rapid removal of ice is the rise in sea level due to the melting ice sheets. Large ice sheets adjacent to the oceans are undercut by rising sea levels because the coldest seawater can be is about −1.8°C while the base of the ice sheet is usually colder than −30°C, so it is like putting hot water under a tub of ice cream. Undercutting of the ice sheet leads to more melting and ice calving into the ocean. This in turn increases sea levels, which causes even more undercutting. This sea-level feedback mechanism can be extremely rapid. Once the ice sheets are in full retreat the other feedback mechanisms discussed here are thrown into reverse.

Orbital forcing and the African climate

Orbital forcing has an obvious impact on high-latitude climates, drives glacial–interglacial cycles, and influences global climate transitions such as the intensification of northern hemisphere glaciation. It also has a huge influence in the tropics, particularly through precession and its effect on seasonality and, thus, rainfall. There is a growing body of evidence for precession-forcing of moisture availability in the tropics, in Africa, Amazonia, and Asia. The precessional control on tropical moisture has also been illustrated by climate modelling. Amy Clement, a climate modeller at the University of Miami, used a complex climate model, usually used to model future climate changes, to switch between extremes of precession to see what effect it had on climate. She and her colleagues found that a 180° shift in precession could change annual precipitation in the tropics by at least 200 mm/year, and cause a significant shift in seasonality. This is on the same order of magnitude as switching between a full glacial and interglacial period. In contrast, precession has almost no influence on global or regional temperatures.

The mechanism by which precession changes the hydrological cycle in the tropics is very similar to that which causes the monsoons. In the tropics, the Sun's energy is most intense when it is overhead. This heats up the land or sea and thus warms the air above. This warm, moist air rises, leaving an area of low pressure beneath it, which sucks in air from the surrounding area. This suction results in the trade winds, which can travel from high latitudes to this area of rising air. As the winds come from both the northern and southern hemisphere, this area is known as the Intertropical Convergence Zone (ITCZ). As the air at the ITCZ rises it forms huge towering clouds and produces large amounts of rain. The ITCZ moves north and south with the seasons as the position of the most intense sunlight shifts up and down across the equator. It is also strongly influenced by the position of the continents. This is because the land heats up faster and to a greater extent than the ocean, and thus it can pull the ITCZ even further north or south during the respective seasons.

Eccentricity affects the Sun–Earth distance as the Earth circles the Sun. Precession affects which time of year coincides with the closest Sun–Earth distance, and thus the amount of insolation received during each season. For example, in the northern hemisphere summer the tropics and subtropics heat up as the Sun steadily moves from overhead at the equator to overhead at the Tropic of Cancer. At the maximum positive precession then, the Sun–Earth distance will be closest when the Sun is overhead at the Tropic of Cancer. This significantly increases the amount of solar energy hitting the subtropics, and thus the amount of convection, strengthening trade winds and rainfall within the ITCZ. Hence the amount of rainfall in the northern hemisphere tropics is greatly increased.

At the same time, on the flip side, the southern hemisphere summer will coincide with the longest Sun–Earth distance; thus the opposite will occur, and rainfall will be greatly reduced in the southern hemisphere tropics. About 21,000 years later the opposite will occur, and the southern hemisphere tropics will have the most intense insolation and thus greatly increased rainfall.

This inverse relationship between the northern and southern hemisphere hydrological cycles can be seen in palaeorecords over the past 10,000 years. North African lakes have steadily been drying out, while the Amazon River discharge has steadily increased over this period of time. We should also remember that this influence of precession increases at the peak of eccentricity every 96,000 years, and is greatest once every 413,000 years when the Earth's orbit around the Sun is at its most elliptical.

In North and East Africa there are excellent records of this precessional forcing of the hydrological cycle over the past 5 million years. First there are the marine dust records for the East Mediterranean which reflect the periodic increase in aridity over the eastern Algerian, Libyan, and western Egyptian lowlands north of the central Saharan watershed. Second, there are complementary dust records from ocean sediment cores in the North Atlantic Ocean adjacent to the West African coast and from the Arabian Sea. Third are the sapropel formations found in marine sediment recovered by deep-sea drilling in the Mediterranean Sea. A sapropel is a dark organic-rich layer found in marine sediments, resulting from a reduction in the oxygen content of the bottom waters. Sapropels are common within Mediterranean marine sediments and are caused by increased Nile River discharge, which changes the circulation of the Mediterranean Sea. The dust records that show aridity, and the sapropel records that show increased rainfall, are inversely related and have a dominant periodicity of 21,000 years—the beat of precession. There is also a growing body of evidence for precessional forcing of East African lakes, which is discussed in detail later in the next section.

We have gone far into the complex world of orbital forcing and seen how precession dominates the climate variability of the tropics, so maybe I should stop here. But although the direct influence of orbital forcing on the African climate seems straightforward, there are additional complexities of precession that need to be considered, as these increased the variability of the climate in East Africa experienced by our ancestors. During any year in the tropics, there are two insolation maxima when the Sun is overhead at the equator—the spring and autumn equinoxes—which, at the moment, occur on 21 March and 21 September. There also two insolation minima when the Sun is overhead at the Tropics of Cancer and Capricorn—summer and winter solstice—which, at the moment, occur on 21 June and 21 December. The magnitude of the maxima and minima, and thus of insolation at equinox and solstice, is, as we have seen, controlled by precession via the changing Earth–Sun distance. In East Africa today, these changes create two rainy seasons, roughly coinciding with the spring and autumn equinoxes—one in March–April, and another in late September–early October. Hence, when precession is increasing insolation of the spring equinox, it is also decreasing insolation during the autumn equinox; therefore one rainy season is getting strong and longer, while the other gets weaker and shorter. So there are two peaks in rainfall for the tropics every half precessional cycle, approximately every 11,500 years. There are also four maximum peaks in seasonality—i.e. the difference between the maximum and minimum solar insolation occurring in any one year in the tropics. This is because the strength and duration of the dry periods between the rainy seasons is determined by insolation at summer and winter solstice. Hence, maximum variation in seasonality in the tropics occurs every 5,500 years.

As I have often argued, it is unsurprising that the tropics are a hotbed of evolution, because even if there were no major climate changes, the whole region undergoes massive hydrological shifts every 5,500, 11,000, and 21,000 years. The key to precessional forcing, however, is how sensitive the region is to either overall changes in yearly rainfall or changes in seasonality, and this determines which periodicity is the dominant control on the environment.

East African Rift System lakes

The sedimentary record of East Africa is rich in lake deposits. This is because the southward propagation of rifting created lake basins along the entire length of the Rift Valley. The map in Figure 32 shows where these lake basins occur, and whether they currently contain a lake. The tectonic movements within the Rift Valley have led to many past lake sediments being exposed by faulting or uplift. They range from huge vertical deposits that, from a distance, look like the white cliffs of Dover, to large white bands sandwiched between fossilized soil horizons.

Because of the dispersed nature of these deposits, the geologist and palaeoclimatologist Martin Trauth at Potsdam University has developed a new twenty-first-century style of fieldwork and very kindly drags me along. He sets up a base camp on the Rift shoulder where the daily temperature range is much smaller than in the Rift Valley. The mobile camps also take advantage of local forest and water sources. This is no safari lodge—it consists of tents and holes in the ground for the toilets—but the food is always wonderful. To get to the sampling sites in the Rift Valley we get up at 6.30 a.m. for breakfast, and as the Sun comes up we board the 'Rift Valley taxi'—the helicopter. Thirty to forty five minutes

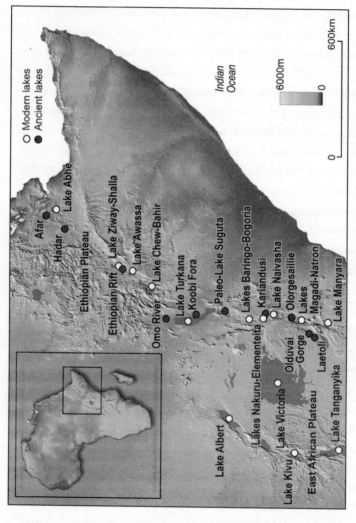

FIGURE 32 East Africa, modern and palaeo lake location map.

later we will be at the first site we have identified. Throughout the day, as soon as we have finished with a site, we can then use the helicopter to move to the next nearest site of interest. When collecting samples and mapping lake and tectonic features, this is an extremely efficient way to run a field campaign in a study area of over 12,000 km^2 that includes sand dunes, rivers, jagged lava flows, and swamps. Otherwise it involves hours and hours in a 4×4 just trying to get to a site, and if there is nothing of interest at that site the whole day is wasted. By around 2.30 p.m., after seven hours' work, and as temperatures in the Rift Valley start to climb over 40°C, we can escape back to the relative cool of the camp to collate and analyse our samples and data.

Using this field method, Martin Trauth and his team have identified and mapped large, deep freshwater lakes that filled the Rift Valley in the past. The key indicator is the presence of pure white, and frequently laminated, or finely layered, sediments made up of diatoms. Diatoms are a major group of photosynthetic algae, found in both freshwater and seawater. They are unicellular, although they can form colonies in the shape of filaments or ribbons, fans, and even stars. The diatom cell is enclosed within a cell wall made of silica (hydrated silicon dioxide) called a frustule. These frustules are unique to each species and show a wide diversity in form, but they are usually bilaterally symmetrical—hence the group name. Diatom species and communities are a useful tool for monitoring environmental conditions, and have been used in seminal studies of water quality and the effects of acid rain. In African lake deposits, the diatom species found can tell us how fresh the water in the lake was, and the approximate depth of the lake. These 'diatomite' sediments are also of commercial value, and in parts of Kenya they are mined, heated in a furnace, and the resultant fine-grained material is shipped all over the world to soft drink manufacturers to filter out any impurities in their drinks. Shallow and more alkaline lakes can also be identified from analysis of lake deposits. These deposits have more mud and sand in the sediment, which would have been washed into the shallower lake, and diatom species that prefer higher alkaline conditions. From our fieldwork we found that deep freshwater lakes were characterized by an area of several hundred square kilometres, water depths in excess of 150 metres, and a neutral pH, and are documented by these pure white and frequently layered diatomites. In contrast, the shallow and more alkaline lakes were typically less than 150 km² in size, had water depths much less than 100 metres (often only a few metres), and dried out episodically. The pH of these lakes is often around 8, but may reach significantly higher values. The corresponding sediments are diatomites with significant amounts of clays and silts. In extreme alkaline lake sediments, we have found silicates such as zeolites, that were generated *in situ* and document chemical weathering of silica volcanic glass due to the extreme pH.

Martin Trauth, myself, and colleagues have compiled a record of these lake occurrences within the Rift Valley. The collation is based on our colleagues' detailed published geological evidence, and our own fieldwork, identifying the appearance of either deep or shallow alkaline lakes within East Africa. To make things easier, we looked at the lakes within seven major basins. These basins, shown in Figure 33, are: Olduvai (Tanzania), Magadi-Natron-Olorgesailie (north Tanzania and south Kenya), Central Kenya Rift (Kenya), Baringo-Bogoria (Kenya), Omo-Turkana-Suguta (north Kenya), Ethiopian Rift (south and central Ethiopia), and Afar (north Ethiopia). Dating of the sediments in East Africa is relatively

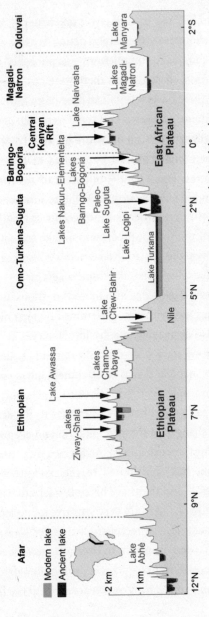

FIGURE 33 East Africa north–south cross section with major modern and palaeo lakes shown.

straightforward because of all the volcanic material, which lends itself well to radioactive argon–argon dating (see Box 3).

The first of these lake compilations was published in the journal *Science* in 2005, showing that there were key periods of time when many of these basins contained deep lakes. Our most recent compilation (Figure 34) shows that there were lake periods 4.6 to 4.4 million years ago, 4 to 3.9 million years ago, 3.5 to 3.3 million years ago, 3.1 to 3 million years ago, 2.7 to 2.5 million years ago, 2 to 1.7 million years ago, 1 to 0.8 million years ago, and 0.2 to 0 million years ago. These occurrences of deep lakes correlate with the 400,000-year component of the eccentricity cycle, and was our first hint of the role of precession in lake formation in East Africa. Moreover, the last four lake periods also correlate with major hominin evolutionary and dispersal events.

Over very long timescales of hundreds of thousands of years, changes in lakes are primarily determined by tectonics, initially creating but also destroying lake basins. However, tectonics also affects conditions in a lake over shorter timescales, through changes to the shape and size of catchment areas and drainage networks, for example. Furthermore, tectonics shapes the morphology of lake basins and hence contributes to the sensitivity of these lakes to changes in the precipitation and evaporation balance. The rifting process in East Africa not only created many lake basins, but has contributed to their sensitivity to small changes in rainfall, and they have become referred to as amplifier lakes. These amplifier lakes are very sensitive to moderate climate change. For example, the reconstructed water level of palaeo-Lake Suguta changed from zero to over 300 metres deep due to a 25 per cent increase in precipitation during the African Humid Period, between 15,000 and 5,000 years ago. On the other hand,

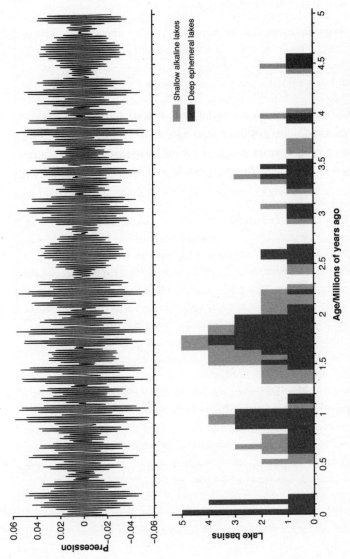

FIGURE 34 Occurrence of shallow and deep lakes over 5 million years compared with the precessional orbital forcing.

Box 3: The dating challenge

To understand the causes of human evolution we need to be able to date the fossils, the hominin artefacts, and the environmental records. This is a considerable challenge, as our ancestors' fossilized bones and artefacts are not directly datable, hence we have to try to date the sediments in which they were found. Key to this is the stratigraphy of the fossil site or location, because very often it is layers above or below the fossil bed that contain material that can be dated. This gives a minimum or maximum age of the fossils. Working out the stratigraphy of a site or location is not easy, especially in East Africa, which has been tectonically very active, so that sediments have been moved and shifted around. Geologists spend many years in the field just piecing together the sequence in which the sediments were laid down to ensure that we can constrain the fossil and artefact ages as much as possible. The hominin fossils found in South Africa are even harder to date, as most of them have been found in cave deposits bereft of datable material. An exception is the Rising Star Cave system just outside Johannesburg where *H. naledi* was discovered and dates have been obtained from the fossil teeth, sediments encasing the fossils and overlying flowstone. Moreover, our knowledge of each fossil site continually improves and so does our ability to date material. An example of this is *H. floresiensis*—the Indonesian cave deposits were initially thought to be 11,000 years old, but after another ten years of detailed, painstaking work this has been re-evaluated and the deposits are now thought to be between 50,000 and 100,000 years old.

There are two primary dating technologies that apply to the study of hominin evolution; palaeomagnetism and radiogenic isotopes (Figure 35). Palaeomagnetism uses the direction of magnetic particles that are found in layers of sediment and compares these to the known shifts in the Earth's magnetic field over time. These shifts include full reversals when the north magnetic pole suddenly becomes the south

continued >

magnetic pole. These seem to occur completely randomly through Earth history and can occur as close as 100,000 years or as far apart as 1.5 million years. You can also use smaller changes in direction of the magnetic field due to magnetic pole wander and variations in magnetic field strength. At the moment the magnetic North Pole is not at the same position as the Earth axis of rotation North Pole, and we know that in the past it has wandered as far south as the US–Canadian boarder. The drawback of palaeomagnetism is that it is not a direct dating method, as you need to know roughly where you are in time before you can interpret the results. This is because what you get is a sequence of normal (north is north) and reversed (north is south) sediments and hence you could be anywhere in the geological record. So some way of tying down roughly where you are in time is essential before palaeo-magnetics can help you refine the date. For example, in a known Miocene sequence of rocks palaeomagnetics was used to date *Sivapithecus*, a genus of extinct primate and probable ancestor to orangutans, to 12.5 million years ago.

Another very useful relative dating method is orbital forcing stratigraphy. This method can be used on long, continuous oceans and lake sediments records. As discussed in Chapter 5, there are regular changes in global climate between glacial and interglacial periods. For the past 1 million years these have occurred about every 100,000 years, and before that they occurred every 41,000 years. So by measuring climate proxies in either lake or ocean cores, which respond to these large changes in global climate, you can count back successive glacial–interglacial cycles and get a chronology for the whole of your record (see Figure 27). This approach can be refined in the tropics, as we know that the tropical climate responds more to precession than to the glacial–interglacial cycles, and so long records that contain the 21,000-year precessional cycle can be used to produce an even higher resolution chronology.

continued >

What we all dream of, however, is sediments that can be dated directly using radiogenic isotopes to give a real date. We also dream that we can date the sediment in which the fossils or artefacts are found, but that rarely happens. The French physicist, Henry Becquerel, discovered the natural radioactive decay of uranium at the turn of the twentieth century. Further work by physicists Ernest Rutherford and Bertram Borden Boltwood indicated that the predictable decay of radioactive elements could be used to keep track of time. Not only did this approach confirm the Earth to be several billion years old, it also provided the earliest empirical timescale for the fossil record. Common chemical elements used in dating include carbon, uranium, potassium, and argon (Figure 35). Which elements researchers use depends on how quickly they decay. For example, radiocarbon (carbon-14 or ^{14}C) dating is extensively used to date organic and carbonate finds less than 60,000 years old, as it has a relatively small half-life (the time it takes for one-half of the atoms in ^{14}C to disintegrate) of about 5,730 years. This has allowed us to document accurately the origins of agriculture, to around 11,000 years ago, and the later occurrence of H. sapiens and the extinction of Neanderthals. Radiocarbon dating does have its drawbacks. First, you can only get a date if you have enough uncontaminated carbon in your sediment. Second, the amount of ^{14}C in the environment

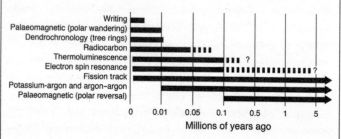

FIGURE 35 Age range of different dating methods used in the study of human evolution.

continued >

varies due to its changing production rates in the upper atmosphere and the release of old 'dead' carbon with no ^{14}C from the oceans or from old sedimentary rocks like limestone and chalk. So huge efforts have been made to calibrate radiocarbon years to calendar or actual years so that we can covert between the two.

Earlier on I complained that trying to piece together the geology of East Africa is difficult because of the tectonics, which has resulted in massive changes in the landscape. But this also means that there is a large amount of volcanic ash or lava interleaved with the lake and river deposits in which fossils and artefacts are found. These ashes and lavas contain potassium-rich minerals such as potassium feldspars, which are essential for potassium–argon and the newer argon–argon dating. Of the naturally occurring isotopes of potassium, ^{40}K is radioactive and decays into argon (^{40}Ar). This decay is at a known rate, as ^{40}K has a half-life of 1.25 billion years, so that the ratio of ^{40}K to ^{40}Ar in minerals is always proportional to the time elapsed since the mineral was formed. The problem is that at room temperature potassium is a solid while argon is a gas, making the isotopic comparison difficult. To get round this, samples are placed in a nuclear reactor to convert the stable form of potassium (^{39}K) into the radioactive ^{39}Ar. It is then possible to measure the ratio of ^{40}Ar–^{39}Ar and thus to calculate the age of the sample. Argon–argon dating has been used successfully to date the majority of the fossil sites in East Africa due to the ubiquitous nature of volcanic sediments in the region. Argon–argon was used to date hominin remains at Dmanisi in Georgia, giving a date of 1.8 million years ago. It has also been used to date the fluvial sandstone layer from which the oldest known fossils of H. floresiensis were excavated in 2014, suggesting that this hominin existed around 700,000 years ago.

There are other less common dating methods such as thermoluminescence and electron spin resonance (ESR), which are used when other datable material is not present. These measure the quantity of electrons that get trapped inside a rock or an artefact over time. First developed in

continued >

the 1950s, thermoluminescence works on the basis that crystals, such as quartz, trap electrical charges (electrons) at a known rate over time. If a crystal, for example a flint blade, is heated, these electrons are liberated, emitting a measurable amount of light. This allows researchers to determine when an object was last heated. What is actually determined is the amount of elapsed time since the sample had previously been exposed to high temperatures. It is used to date volcanic materials and meteorite impact craters, but can also directly date past human activity, such as when ceramics or flint materials were heated to improve strength and durability. ESR also measures electrons trapped in the lattice, but unlike thermoluminescence it does not destroy the sample being tested, allowing samples to be dated multiple times. ESR mostly tests calcium carbonate in limestone, coral, fossil teeth, molluscs, egg shells, quartz, and flint. Both methods are typically used to date samples from the last 300,000 years, though theoretically they could be used on much older samples.

as hydrological modelling suggests, large water bodies buffer rapid shifts in climate due to their delayed response to changes in the precipitation–evaporation balance. Thus, theoretically, lakes can be very quick to form, but their influence on the local climate creates an inertia that resists their removal or disappearance.

Subsequently other groups have supported our work, finding more evidence for the role of precession. For example, the geochronologist and Chelsea fan Alan Deino (Berkeley Geochronology Centre) and palaeoanthropologist John Kingston (University of Michigan) have found that the major lake period found in the Baringo Basin in the Central Kenyan Rift, between 2.7 and 2.55 million years ago, actually consisted of five tick diatomite deposits representing distinct palaeo-lake phases separated by a the precessional cyclicity of ~21,000 years. Clayton Magill and colleagues at

Pennsylvania State University have found biomarker stable carbon isotope evidence (see Box 1) in Olduvai lake sediment of precessional forced variations between open C_4 grasslands and C_3 forest between 1.9 and 1.8 million years ago. There is also evidence for precessional forcing of the lake phase 1.9–1.7 million years ago identified in the KBS Member of the Koobi Fora Formation in the North East Turkana Basin in Kenya. During the same period, an oxygen isotope record from the Buffalo Cave flowstone in the Makapansgat Valley, Limpopo Province, South Africa, shows clear evidence of precessionally forced changes in rainfall. The occurrences of these environmental changes are in phase with increased freshwater discharge into the Mediterranean Sea, shown by the sapropel records, and out of phase with the dust transport records in sediments from the Arabian Sea, tropical North Atlantic Ocean, and Mediterranean Sea.

In summary, then, the palaeoclimate records from East Africa and surrounding regions show that there were distinct periods when the local climate became highly variable. These periods coincide with the approximately 400,000 years maxima in eccentricity. During each of the periods, the climate of East Africa cycled between extremely wet periods, when the whole of the Rift Valley was full of deep freshwater lakes, to hyper-arid conditions when huge amounts of dust were blown into the surrounding oceans. This extreme climate variability varied on a 20,000-year frequency, suggesting that precession forces the local climate to switch between extremely wet and dry phases. These periods of highly variable climate seem to last only 100,000 years. In between there are long periods of time when there seems to have been very little change in the East African climate, and no lakes, apart from large permanent ones such as Lake Turkana.

These short periods of extreme climate variability seem to correlate with key periods of hominin evolution, such as the first appearance of *A. ramidus* (4.4 million years ago); of the genus *Homo* (2.8 million years ago) and the genus *Paranthropus* (2.5 million years ago); of *P. robustus, H. erectus,* and *H. ergaster* (1.8 million years ago); of *Homo heidelbergensis* (700,000 years ago); and of *H. sapiens* (200,000 years ago).

The new understanding of the hominin fossil record, the tectonic history of East Africa and short-term extreme variations in the local environment mean we can now examine theories of human evolution in a more critical way. In the next Chapter we examine the existing theories and see whether there is a way that a new view of what caused our evolution is starting to emerge from all this new data.

7

African Climate Pulses

In previous chapters we have seen the huge changes to the landscape of East Africa wrought by tectonics, celestial mechanics, and climate change. The hominin fossil record gives us brief glimpses of our different ancestors and when they first and last appeared in Africa. In this chapter we will look briefly at the different theories that have been put forward in which a changing environment is the key control on hominin evolution. My colleagues and I have developed a new synthesis of the palaeoenvironmental data from East Africa that allows us to bring together all the different theories of human evolution into one framework. It seems as if no single theory may be right, but I will instead argue that all of them may be right.

Theories of evolution

As we have seen, environmental pressures have long been assumed to play a key role in hominin speciation and adaptation. A number of iconic theories have been put forward to frame and develop the discussion of hominin evolution. In Table 2, I have tried to put these theories into the context of overarching theories of evolution.

There are two main schools of thought when it comes to overall evolutionary theory. The first is *phylogenetic gradualism*, based on the idea of evolution presented in Darwin's *Origin of Species*, which suggests that most speciation is due to one species slowly, uniformly, and gradually evolving into a new and distinct species. However, most scientists would agree that there is little or no evidence for this uniform rate of evolution, and evolutionary biologist Richard Dawkins argues that Charles Darwin himself was not a constant-rate gradualist. For example, in the first edition of *On the Origin of Species*, Darwin stated that 'Species of different genera and classes have not changed at the same rate, or in the same degree'. Hence, we must differentiate between constant and variable phylogenetic gradualism.

The second school of thought that developed from Darwin's original ideas was first proposed in 1972 by palaeontologists Niles Eldredge and Stephen Jay Gould—they called their new theory *punctuated equilibrium*. They propose that once species appear in the fossil record they will become stable, showing little evolutionary change for most of their geological history—a state they called *stasis*. When a rare and geologically rapid event occurs, then significant evolutionary change occurs through branching speciation called cladogenesis. Cladogenesis is the process by which a species splits into two distinct species, rather than one species gradually transforming into another. When we look at the whole geological record, life is dominated by extinctions and speciations due to the five large mass extinctions, the last of which ended the rule of the dinosaurs and ushered in the age of mammals. So, with a very large timescale, punctuated equilibrium seems a good

TABLE 2. Theories of evolution and how they correspond to the many postulated theories of early human evolution

			CLIMATE STRESS			
			None	Long-term state change	Threshold event	Variability
EVOLUTIONARY FORM	Phyletic gradualism	Constant	Red Queen	*Allopatic speciation*		
		Variable	Red Queen	Turnover pulse hypothesis Savannah hypothesis	*Allopatic speciation*	Variability selection hypothesis Pulsed climate variability hypothesis
	Punctuated equilibrium		Court Jester (e.g. impact event)	Aridity hypothesis	Court Jester *Allopatic speciation*	Pulsed climate variability hypothesis

fit with our knowledge of the fossil record. But with smaller time-scales, such as those we are investigating for hominin evolution, punctuated equilibrium may not be so applicable. Of course, we must remember that the split between phylogenetic gradualism and punctuated equilibrium is artificial, but it does provide a starting point for discussing theories of early human evolution.

Theories of early human evolution

Various evolutionary theories have sought to explain the evolution of bipedalism, encephalization (brain expansion), and our ancestors' dispersals out of Africa in terms of environmental pressures. The first key environmental theory to explain bipedalism was the *savannah hypothesis*, which suggests that as grasslands expanded, hominins were forced to descend from the trees and adapt to life on the savannah, which was facilitated by walking erect on two feet. The savannah hypothesis has a long history and has influenced the debate about early human evolution for over 200 years. It was first suggested in 1809 by the French naturalist, Jean Baptiste Lamarck, in his book *Philosophie zoologique*:

> If the individuals of which I speak were impelled by the desire to command a large and distant view, and hence endeavoured to stand upright, and continually adopted that habit from generation to generation, there is again no doubt that their feet would gradually acquire a shape suitable for supporting them in an erect attitude. (Lamarck 1809, 170)

This view was echoed sixty-two years later by Charles Darwin, who also envisioned primeval humanity leaving the forest and adapting to open plains. In his book *Descent of Man*, he wrote:

> As soon as some ancient member in the great series of the Primates came, owing to a change in its manner of procuring subsistence, or to a change in the conditions of its native country, to live somewhat less on trees and more on the ground, its manner of progression would have been modified; and in this case it would have had to become either more strictly quadrupedal or bipedal. (Darwin 1871, 140–1)

The savannah hypothesis led to the knuckle-walking hypothesis, which states that human ancestors used quadrupedal locomotion on the savannah, as a precursor to bipedal locomotion. But, as discussed at the end of Chapter 4, the fossil record suggests that our ancestors did not knuckle-walk—instead, they retained long arms needed for tree-climbing at the same time as they developed bipedalism. For example, *Ardipithecus ramidus*, when it existed 4.4 million years ago in Afar in Ethiopia, was clearly able to travel fairly long distances using its bipedal adaptation, but retained the ability to climb trees that would have been essential for food gathering and escaping the large numbers of predators in East Africa. We also now know much more about the development of the East African palaeoenvironment and the slow fragmentation of the vegetation, as opposed to a simple binary switch between forest and savannah.

I can easily understand why scientists working in East Africa could have seen the modern landscape and assumed it was a binary system. This is because as you fly up the Rift Valley there are vast areas that look just like savannah, with mainly grass and a few large acacia trees scattered around the landscape, with some isolated areas of forest. But this landscape is 1,000 metres above sea level, with an annual rainfall that could easily support open forest. The reason it is grassland is because of deforestation and constant grazing by cattle, goats, and camels. The worst offenders for this type of deforestation are the goats, which can climb small trees and strip plants instead of nibbling them, preventing the plants from regrowing. Therefore much of the present-day landscape of East Africa is the result of human activity, making it much harder for us to envisage the past. The current predominant view of the origins of bipedalism is that that it had an arboreal origin. Our ancestors learnt to stand upright in the trees using their forelimbs for support. Bipedalism was a response to the need to move greater distances as efficiently as possible between different food sources as the forest cover started to fragment and change.

The savannah hypothesis was refined as the *aridity hypothesis*, which suggested that the long-term trend towards increased aridity and the expansion of the savannah was a major driver of hominin evolution, particularly the evolution of *Homo* and encephalization. This idea was encapsulated in a paper published in *Science* in 1995 by the palaeoceanographer Peter deMenocal of Columbia University. He used dust records in marine cores off the coasts of the Sahara and Ethiopia to look at changes in aridity on the African continent, and suggested that there were three distinct periods when aridification of East Africa accelerated, related to the major thresholds in the global climate system discussed in Chapter 5:

> Major steps in the evolution of African hominids and other vertebrates are coincident with shifts to more arid, open conditions near 2.8 Ma, 1.7 Ma, and 1.0 Ma, suggesting that some Pliocene (Plio)- Pleistocene speciation events may have been climatically mediated. (deMenocal 1995, 59)

The aridity hypothesis was probably the first of the early human evolution theories that embraced the evolutionary pattern of punctuated equilibrium. For the next decade, this neat and straightforward theory became the textbook explanation for human evolution, until it was challenged.

Another set of theories focused on how different types of species react differently to environmental stress. The *turnover pulse hypothesis*, suggested by Elisabeth Vrba at Yale University, was originally developed in 1988 to explain discrete patterns in ungulate (hoofed animal) speciation, and suggests that acute climate shifts drove adaptation and speciation. She recognized that environmentally induced extinctions hurt specialist species more than generalist species. Hence, when there is environmental disruption the generalists will tend to thrive by utilizing new environmental opportunities and by moving elsewhere to take advantage of other areas that have lost specialist species. The specialists will experience more extinctions, and an increased speciation rate within their group. This would lead to more rapid evolution in isolated areas—referred to as allopatric speciation (Figure 36)—whereas the generalists will become more spread out.

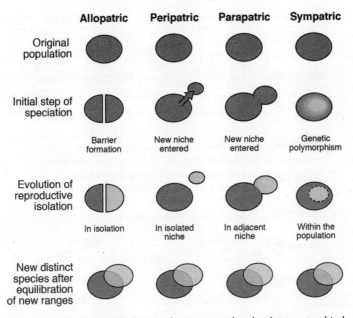

FIGURE 36 Different modes of potential speciation with and without geographical or niche separation.

Rick Potts, the director of the Smithsonian Institution Museum of Natural History's Human Origins Program, later refined Elisabeth Vrba's theory and coined the *variability selection hypothesis*. This theory advocates the role of environmental unpredictability in selecting for behavioural or ecological flexibility. It develops the original turnover pulse hypothesis, but instead splits species in terms of their differing ability to adapt and evolve in a more variable and unpredictable environment (Figure 37). The variability selection hypothesis originally emphasized the long-term trends towards a drier and more variable climate. However, it struggled to explain the recent palaeoanthropological evidence

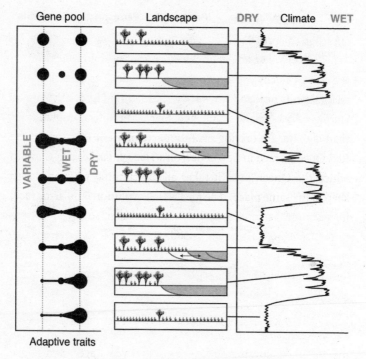

FIGURE 37 Variability selection hypothesis of early human evolution.

suggesting a pulsed/threshold nature to hominin speciation and migration events. So, over the past few years, Potts has revised the variability selection hypothesis so that the selection for more flexible adaptations occurs during the climatically variable periods of time.

More recently, it has been suggested that periods of climate stability may be equally important in driving human evolution, dispersal, and technological innovation. Relatively long periods of climate stability could produce the scenario predicted by the *Red*

Queen hypothesis, or *sympatric evolution* due to sexual selection, discussed in Chapter 8.

Leigh van Valen, an American evolutionary biologist, proposed the Red Queen hypothesis in 1973, which suggests that continued adaptation is needed for a species to maintain its relative fitness among co-evolving systems, and that biotic interactions, rather than climate, are driving evolutionary forces. It is based on the Red Queen's race in Lewis Carroll's *Through the Looking-Glass*, in which the Queen says 'It takes all the running you can do, to keep in the same place.' The idea is that evolution is an arms race; for example, as antelopes evolve to run faster, so cheetahs have to get faster too. However, for this to occur it is reasonable to assume a relatively highly productive environment, so that competition, rather than resources, forms the dominant control. At Koobi Fora (northern Kenya) there is evidence for the presence of multiple hominin species, including *Paranthropus boisei*, *Homo erectus*, *H. habilis*, and *H. rudolfensis*, attributed to the period of maximal lake coverage (~1.9–1.8 million years ago), and hence the highest availability of resources. It might be postulated that these hominins evolved as a result of competition with each other and other animals. It is interesting to note that, at the end of this period, only one species survives—*Homo erectus*. The antithesis to the Red Queen hypothesis is the *Court Jester hypothesis*, coined by the palaeontologist Anthony Barnosky at Berkeley University. He suggested that abiotic forces such as tectonics and meteorite impacts, rather than biotic competition between species, are the major drivers behind the process of evolution.

Finally, a direct development of the variability selection hypothesis, which incorporates the role of both stability and instability and challenges the aridity hypothesis, is the *pulsed climate variability framework*. This theory came out of the original palaeo-lake work of Martin Trauth, myself, and our colleagues, and highlights the role of short periods of extreme climate variability in East Africa in driving hominin evolution. This framework is discussed in the next section, along with how the other evolutionary theories may be applied, given the new environmental context. Instead of any one of these theories being the right one, they could all play an important part in the story of early human evolution.

Pulsed climate variability

Our pulsed climate variability framework suggests that there were periods of extreme climate variability every 400,000 or 800,000 years, driven by the eccentricity maxima, when lakes grew and filled much of the Rift Valley, and then disappeared on an approximately 20,000-year precessional timescale. These periods of highly variable climate in East Africa seem to coincide with the majority of the first and last appearances of hominin species over the past 5 million years. A period of highly variable climate between 4.6 and 4.4 million years ago coincides with the ages of *A. ramidus* discovered at As Duma in the Afar region of Ethiopia. Another highly variable climate period between 3.4 and 3.2 million years ago coincides with the record of *A. afarensis* at Hadar, Ethiopia, covering a period from about 3.4 to 3 million years ago. Between 2.7 to 2.5 million years ago there is another period to which the five diatomite deposits of the Baringo Basin have been identified. These correspond to the peak in precession and the age, about 2.8 million years ago, to which the first *Homo* species have been identified, as well as to the appearance of the *Paranthropus* genus, 2.5 million years ago. The largest and most prolonged period with variable climate is between 2.1 and 1.7 million years ago. This coincides with distinct variants of *H. rudolfensis* and *H. habilis* identified by Susan Anton of New York University and her colleagues, at 2.09 million years ago, the first appearance of *H. erectus* 1.9 million years ago, and the first dispersal of hominins outside Africa about 1.8 million years ago. Then, 1.1 to 0.9 million years ago, another highly variable period coincides with the last appearance of *H. erectus* in East Africa and the second dispersal of hominins out of Africa. Recent modelling work by Rick Potts and Tyler Faith at the University of Queensland suggests that these correlations are not down to sampling bias or coincidence.

Fundamental to understanding what evolutionary mechanisms could have applied to hominins in East Africa are the speed and form of the transitions between lakes appearing and disappearing from the landscape. At first, it may appear that orbitally forced climatic oscillations may be too long-term to have created rapid changes in lake occurrence. However, this does not take account of the sinusoidal nature of orbital forcing or the threshold nature of the African lake systems. All orbital parameters are sinusoidal, which means that there are periods of little or no change followed by periods of large change. For example, the sinusoidal precessional forcing at the equator consists of periods of less than 2,000 years, during which 60 per cent of total variation in daily insolation and seasonality occurs. These are followed by ~8,000 years when relatively little change in daily insolation occurs. Hence, precession does not result in smooth forcing, but rather produces a combination of brief periods of strong forcing and long periods of relatively weak forcing. If this is combined with the idea that many of the East African lakes are amplifier lakes that respond very quickly to a small increase or decrease in the precipitation–evaporation balance, then it is relatively easy to envisage threshold responses of the landscape to precessional forcing.

Katie Wilson, my former doctoral student at University College London, showed that the diatomite deposits in the Baringo Basin do indeed suggest that the lakes appear rapidly, remain part of the landscape for thousands of years, then disappear in a highly variable and erratic way. In fact, the absence of shallow-water diatom species at the start of many palaeo-lake deposits suggests that the lakes could have appeared in only a few hundred years. Figure 38 shows a compilation of what a generic extreme wet–dry cycle may have looked like, with a threshold at the beginning of the wet phase and a prolonged highly variable period at the end of the wet phase. There would be four or five of these cycles during each of the periods of extreme climate variability.

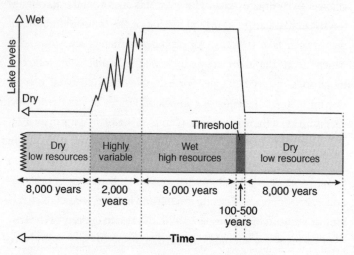

FIGURE 38 Pulse climate variability framework.

The different speeds at which the lakes appeared and disappeared is also consistent with the idea of a bifurcated relationship between climate and lake presence. Figure 39 shows that precipitation needs to increase significantly before lake growth can initiate, but once it has, there are some key feedbacks which accelerate the expansion of the lakes. The most important is the change in the local climate due to increased moisture in the atmosphere. Hence, increased local relative humidity reduces the amount of evaporation occurring, thus increasing the moisture in the atmosphere. When lakes become more established, the general increase in moisture changes the vegetation and more bushes and trees appear, subsequently increasing the evapo-transpiration and further increasing the moisture in the atmosphere. These same feedbacks also resist the drying out of the lake when precession starts to reduce overall rainfall. This leads to a period of up to 2,000 years when the lake expands and contracts, until finally there is not enough moisture in the region to sustain any sort of lake.

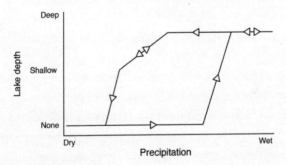

FIGURE 39 Relationship between growth and shrinkage of large lakes in the African Rift Valley. The bifurcation is caused by the evaporation from the lake making the surrounding region wetter and thus resilient to initial reductions of overall precipitation.

More recent evidence of this sort of lake behaviour has been found by palaeoliminologist Annett Junginger, now at the University of Tübingen in Germany, and Martin Trauth. Between them they compiled a large number of radiocarbon dates (see Box 3) for palaeo-Lake Suguta in northern Kenya (see Figures 32 and 33). These show that the lake appeared rapidly 14,700 years ago, achieving a depth of 300 metres in less than a few hundred years. The lake survived until about 7,000 years ago, and then took about 2,000 years finally to dry up completely.

It should be stressed, however, that this pulsed climate variability framework, whereby eccentricity-modulated precession is the main influence, only applies up to 800,000 years ago. After this time, as mentioned in Chapter 5, the Early–Middle Pleistocene transition occurred, intensifying and prolonging the glacial–interglacial climate cycles, which thus have an increasing influence on tropical climates. Hence, over the past 800,000 years the climate of the tropics became more complicated and fragmented as it was influenced both by localized influences of orbital forcing, and the ever stronger global influence of the glacial–interglacial cycles.

Human evolution within the new framework

The big problem with most previous theories of human evolution is that they have no timescale over which they operate. This is important because a slow environmental change over one hundred generations is going to have little or no effect on the evolution of a species. However, the drying up of a lake over a few generations will have an effect on the survival of individuals, and thus potentially drive evolution. It seems harsh, but evolutionary change is equally about those individuals and their traits which do not survive as it is about those that make it through. So the pulsed climate variability framework provides a temporal framework for key periods of time when we know that new species of hominin evolved, and so it can help us to conceptualize the different theories of early human evolution. Figure 40 shows how the turnover pulse hypothesis would operate through one of these extreme climate cycles.

The palaeontologist and zoologist, Elisabeth Vrba, suggested that environmental changes would affect specialist and generalist species differently. During dry periods, the extinction rates of specialists would go up as they would struggle to find food, having lost their environmental niche and their competitive advantage. In contrast, the generalists would have a lower extinction rate, as they would be better adapted to finding resources during stressful periods (Figure 40). Speciation would be much higher among the specialists during dry periods, as they adapt to new habitats. In contrast, during the wet periods, and to a lesser extent the highly variable periods, generalist species would suffer as specialists would have a lot more niches to fill and thus would outcompete the generalists. Figure 41 illustrates possible changes that could have occurred due to the aridity hypothesis, which suggests that speciation mainly occurs during periods of dryness with low resources. Extinctions would have occurred during both wet and dry periods, but for different reasons. They would have occurred during dry periods as some species would not have been able to adapt to the harsh conditions, which conversely drives speciation. There would have also been extinctions during the wet period as some of the species that adapted to the harsh dry conditions may not have retained the ability to cope with the wet plentiful periods, or may have been outcompeted by those more able to take advantage of a high-resource environment.

Turnover pulse hypothesis

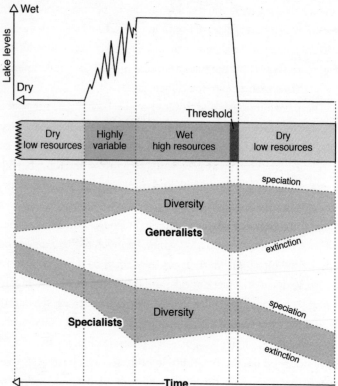

FIGURE 40 An interpretation of how the turnover pulse hypothesis could be placed within the pulsed climate variability framework.

The variability selection hypothesis developed from the original turnover pulse hypothesis, but instead splits species into their varying ability to adapt and evolve in a more variable and unpredictable environment. Hence, generalists undergo more extinction, and specialists more speciation, during the highly

Aridity hypothesis

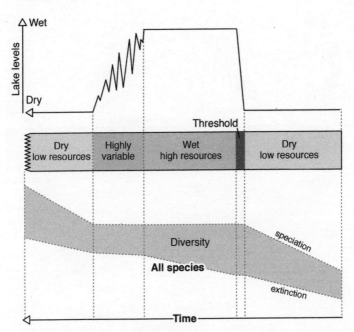

FIGURE 41 An interpretation of how the aridity hypothesis could be placed within the pulsed climate variability framework.

variable climate period in between the long wet and arid phases. In contrast, the Red Queen hypothesis suggests that continued adaptation is needed in order to keep up with other species which are also evolving. It is therefore reasonable to assume that a relatively high-energy environment would provide more resources and therefore more energy in the system to enable interspecies competition.

Finally, the visual example in Figure 42 illustrating allopatric evolution suggests that by geographically isolating populations,

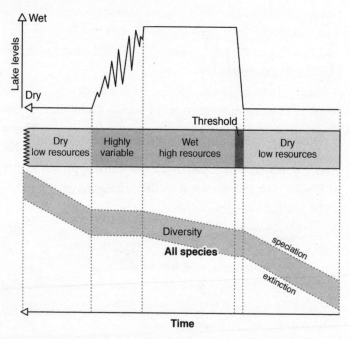

Allopatric speciation hypothesis

FIGURE 42 An interpretation of how the allopatric speciation hypothesis could be placed within the pulsed climate variability framework.

each can then evolve independently with slightly different environmental stresses. In the Rift Valley, during the extreme dry periods, north–south and east–west migration is very difficult, so it would have created isolated populations. The same is true of extreme wet periods, because when lakes completely fill the rift basins migration north–south and east–west would again be difficult, also creating isolated populations. Only during the high-variability period and the threshold change would it be possible to move up and down and across the Rift Valley easily. Recent evidence from

Katie Wilson's work suggests that there were millennial-scale fluctuations in lake levels during the extreme wet periods, probably driven by the northern ice sheet collapse events. This may have made movement between populations possible during the wet phases, limiting the isolation.

The illustrations shown in this chapter are just my interpretations of how the major theories of human evolution could be placed within the pulsed climate variability framework. This is my attempt to place a temporal framework on evolutionary theories. I would encourage you to use this visual approach to play with your own interpretation of how changing environments would interact with different theories of early human evolution. We should remember, however, that there are long periods of stability ranging from 200,000 to 400,000 years between the high-variability climate periods in which evolutionary events could have occurred. Moreover, as we will examine in Chapter 8, the evolution of our brain may have also helped drive our evolution independently of environmental change.

Hominin dispersals

The periodic hominin dispersal events also seem to correlate with these periods of high climate variability. Dispersal events seem to occur at about 3 million years ago within Africa, and at 2.1–1.9, 0.9, 0.6, and after 0.1 million years ago out of Africa (Figure 8). It has been suggested that both lake presence and absence could be associated with these dispersal events. For example, when the lake basins are dry they become 'hyper-arid'; thus the Rift Valley becomes uninhabitable, and hominin populations would have been forced to migrate to the north and south. However, severe lack of resources would mean there was only a small and possibly shrinking population that could migrate. The absence of lakes may have facilitated allopatric speciation in key refugia such as Turkana, which may have remained wet. Dispersal is thus more likely to have occurred when the basins were completely filled with water, as there would have been limited space for the hominin populations on the tree-covered Rift shoulders and river flood plains. The wet conditions could have been more conducive to dispersal because hominin populations could expand due to the availability of water and food, and could follow the Nile tributaries northward and through a green Levant region. There is some clear evidence that when it is wet in East Africa it is also wet in the Levant and the Middle East, opening up multiple corridors out of Africa and into the East. So the occurrence of deep freshwater lakes would have forced expanding hominin populations both northwards and southwards, generating a pumping effect, pushing them out of East Africa towards the Ethiopian highlands and the Sinai Peninsula, or into Southern Africa, with each successive precessional cycle. The recent finding that the wet periods were also variable suggests that there could have been small expansions and contractions of the Rift Valley lakes on a millennial scale, enhancing this pumping effect of hominins out of Africa.

The pulsed climate variability framework therefore takes the latest palaeoclimate understanding of East Africa and provides a framework within which to understand the causes of early human evolution. There seem to have been periods of highly variable climate in East Africa, occurring every 400,000 or 800,000 years. Within these periods the climate of East Africa and the Levant seems to vary from extremely dry, with large quantities of dust blowing into the adjacent oceans, to extremely wet, when the Rift Valley basins were filled with deep freshwater lakes. These variations seem to have been driven by precession. The dry periods may have lasted up to 8,000 years, and then lakes appeared on the landscape very rapidly, maybe within a few hundred years. The wet periods also lasted up to 8,000 years, and there is evidence that the lake levels fluctuated up and down, driven by millennial-scale events probably originating in the North Atlantic Ocean. At the end of the wet period, the lakes took nearly 2,000 years finally to dry out, during which time lake levels were highly variable. This means that there were environmental fluctuations on the correct timescale to influence evolution. Different species or, at the very least, different emerging traits within a species, could have evolved through various mechanisms, including the turnover pulse hypothesis, aridity hypothesis, variability selection hypothesis, or allopatric speciation. I argue that this is exemplified by the case of *Homo erectus*, which exhibited changes in life history (shortened interbirth intervals, delayed development), pelvic morphology, body size and dimorphism, a shoulder morphology that enabled projectile use, adaptation to long-distance running, ecological flexibility, social behaviour, and a new stone tool kit allowing the processing of food. Each one of these traits could have been forced by a different evolutionary mechanism operating at a different part of the environmental cycle. So what is emerging from our new understanding of the hominin fossil record and the palaeoenvironment of East Africa is a much more complicated and rich picture of how our species evolved.

In Chapter 8, we will consider how a large brain could have helped our ancestors cope with these rapid environmental changes.

8

The Social Brain

The previous chapters give us some insight into the rapid environmental changes that occurred in East Africa coincident with the emergence and disappearance of hominin species. It is easy to imagine that many of these environmental changes would have created stress due to reduced access to food and water, and would have created positive selective advantages for those ancestors whose particular traits enabled them to maintain access to essential resources. The fossil record seems to suggest that there were many different hominins roaming throughout Africa at the same time, providing fertile ground for evolution. At critical periods, when populations of different hominin species were geographically together—such as on the shores of Lake Turkana 1.8 million years ago—we can imagine species-to-species competition and possibly interbreeding. This species-to-species competition may simply have been about winning or losing access to food and water. Or it may have led to direct conflict. The seminal work of the renowned primatologist Jane Goodall on chimpanzees shows that they can sustain long-term intraspecies warfare until one group of male chimpanzees is wiped out. There may have also been interbreeding between hominin species—for example, recent genetic studies show that there was some interbreeding between *H. sapiens* and *H. neanderthalensis*. There is, however, another evolutionary driving force that needs to be examined—sexual selection. In this chapter I examine how this may have driven the evolution of a larger brain through the need to control large social groups.

What good is a large brain?

It may be possible that a large brain became a key characteristic in sexual selection, when both males and females actively selected partners with traits associated with a bigger brain. This would have led to sympatric evolution (Figure 36), when new species emerge from within the original population and then disperse. The *social brain hypothesis* suggests that enhanced cognitive ability would provide the ability strongly to influence groups or tribes of hominins, and hence control the distribution of resources, and thus it would be strongly selected for, driving brain expansion in hominins.

Originally, large brains were thought to be essential to enable the making of stone tools, and this is why *H. habilis* was thought to be the start of the *Homo* genus. We now know that many animals make and use tools, and that stone tools originated much earlier than *Homo*. So what was the large brain used for if it was not essential for tool-making? Anthropologist Robin Dunbar and colleagues at Oxford University argue that it takes a huge amount of cognitive ability to exist in large social groups. Our position in the group, our access and ability to get mates and resources, rely on how good we are at playing the social game. As the groups get larger, so the computational power needed to keep up with the interconnections grows exponentially. For example, if you are talking to your best friend about the weather, this is a simple linear conversation with little brainpower needed. If you now talk to your best friend about their mother you need to start to think. First, what is their relationship to their mother? Has it been good or bad in the past, and what is it like now? What is your relationship like with their mother, and how does this relate to your relationship with your best friend? Let's take it up a notch—how about a simple question to your best friend about how their mother is getting on with her husband, who is in fact your best friend's stepfather. This is a simple example with someone you know well. Imagine what happens when it is someone you know less well and who you want to impress.

I have to say I am very mean when I illustrate this to my first-year undergraduates—usually by asking one who has been repeatedly late to my lectures a set of simple questions. Remember, this person is answering a professor in control of a class of 130 students. The first question, 'How are you?', usually gets a simple answer and sometimes a bit of humour such as, 'I was better five minutes ago.' The second question, 'Do you like my class?', is a difficult one because they want to appease me, but also do not want to be seen as weak or cowering before me. The third question is, 'Do you like this person sitting in the second row?' (I point to that person). They now have to decide what their relationship is to me, to the person in the second row, and what they wish to tell me and the other 128 students. Because in complex social situations the truth is extremely fluid, the answer they give is usually one to make them look good or simply extricate themselves from the nightmare situation I have put them in. The fourth question is, 'Do you think they would go out with your friend sitting next to you?' This one completely throws them as they have to work out their answer based on their relationship with me, their friend, the person sitting on the second row, and the other 127 students. Nine times out of ten the student will use humour to get out of the situation. But remember, all these stressful social replies have to be worked out in milliseconds so that the conversation continues.

So social groups are complex, with high stress levels, because the rewards are high. Hence, our huge brain is developed to keep track of rapidly changing relationships. My undergraduates come to university thinking they are extremely smart as they can do differential equations and understand the use of split infinitives. But I point out to them that almost anyone walking down the street has the capacity to hold the moral and ethical dilemmas of at least five soap operas in their head at any one time, and that is why we have a huge complex brain. And our brain is truly complex because to just understand how the brain maintains our human intellect, we would need to know about the state of all 86 billion neurons and their 100 trillion interconnections, as well as the varying strengths with which they are connected, and the state of more than 1,000 proteins that exist at each connection point. But neurobiologist Steven Rose suggests that even this is not enough, as we would need to know how the connections have evolved over a person's lifetime and even the social context in which they had occurred. In *The New York Times*, the neuroscientist Kenneth Miller suggested it will take 'centuries' just to figure out basic neuronal connectivity. This new understanding of the human brain and its capacity and complexity has also changed our views about computers. In the 1970s and 1980s it was thought that artificial intelligence (AI) was catching up with humans. With all our new knowledge of the human brain and how complex it is, AI is now even further behind than we thought. So, for all the recent hype, the takeover of the world by sentient robots is scientifically even further away than it was when initially envisaged.

However, the invention of computers has caused another problem, which is that many scientists think the human brain operates like a computer. However, Robert Epstein, a psychologist at the American Institute for Behavioral Research and Technology, says this is just shoddy thinking and is holding back our understanding of the human brain. He points out that humans start with senses, reflexes, and learning mechanisms. What we do not start with and never have are: information, data, rules, software, knowledge, lexicons, representations, algorithms, programmes, models, memories, images, processors, subroutines, encoders, decoders, symbols, or buffers—which are key design elements that allow digital computers to behave somewhat intelligently. Robert Epstein argues that the 'information processing' metaphor of human intelligence is completely wrong and is holding back the whole of neuroscience. As he bluntly puts it, 'We are organisms, not computers. Get over it.' The final difference is that computers store exact copies of data that persist for long periods of time, even when the power is switched off, while our brain only maintains our intellect as long as it remains alive.

I would argue that the incredible flexible and complex hominin brain evolved to deal with complex social groups so that an individual could maximize their access to the best resources and the opportunity to mate with the best partners. I also do a small experiment each year with my first-year undergraduates to show how sexual selection for bigger brains may have operated. So every year, for the past twenty years, in my human evolution lectures, I asked my undergraduates questions about how they go about choosing a mate. As you can imagine there is a lot of sniggering. I randomly ask female undergraduates what traits they look for if they are interested in a male partner. The answers are usually humour, intelligence, charm, etc. Now these are great, but I point out that I cannot see how those qualities would have helped our early ancestors on the plains of Africa when faced by a hungry lion. Would they not have preferred a 'man who throws a spear furthest in a pub', which sets off more laughter. This line of questioning has a serious point because, though human societies value good looks and youth, they also highly value humour, intelligence, and individuals with access to large resources. This is because those traits show us who are alphas in society with the ability to control large groups of humans and thus co-opt a greater share of resources for themselves and their family. Extreme examples of this power are top comedians who can fill stadiums and control and entertain 70,000 people, or politicians who, through their rhetoric and charm, convince millions of us to vote for them so they can run our countries.

Group size: the magic number 150

Robin Dunbar has created a magic number: 150. He and his colleagues found that modern humans like communities of about 150 people. This number is found in the population of Neolithic villages (6,500–5,500 BC), villages in the Domesday Book (1085 AD), eighteenth-century English villages, modern hunter-gatherer societies, Anglican church congregations, and Christmas card distribution lists—to name just a few examples. He has also shown that community size in primates is linked to the size of the neocortex region of the brain, where all cognitive processing occurs. If one extrapolates the relation based on the other extant primates then humans with their very large neocortex extend the graph to a community size of 150. This leads Robin jokingly to say that anyone declaring that they have more than 150 friends on Facebook is lying. Through detailed analysis of Twitter communities (those who exchange posts with each other, based on a sample

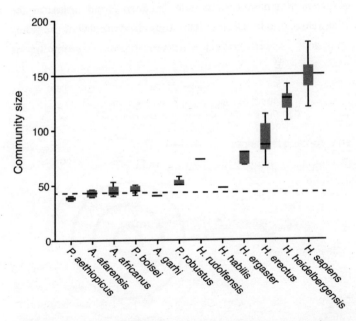

FIGURE 43 Estimated community size of different hominin species over the past four million years. Community size is estimated from converting cranial volumes to neocortex ratios, which have a strong relation to group size in primates. Median shown with 50 per cent and 95 per cent range. Solid line is the major 150 community size found within modern human societies.

of 1.7 million people) and email communities (those that exchange regular emails based on a sample of 10 million people) his team showed that these virtual communities also usually consisted of between 100 and 200 individuals.

This relationship between community size and the size of the neocortex is extremely useful, as it means we have a way of estimating group size in extinct hominins. Hominins such as the *Australopithecines* and early *Homo* have an estimated community or group size of fifty. The big change occurs with *Homo erectus*, when group size jumps up to nearly one hundred; then *Homo heidelbergensis* at 130 and, of course, modern *Homo sapiens* at 150 (Figure 43). Finally, Dunbar has shown that the size of the group in primates directly correlates with the amount of grooming. He

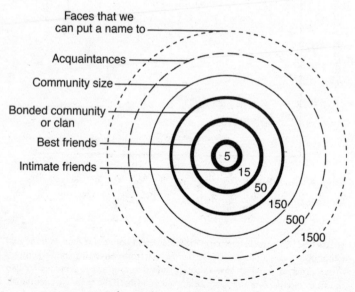

FIGURE 44 Robin Dunbar's circles of human relationships.

argues that living in large groups is extremely stressful, so grooming is a way of lowering individual and group stress. This grooming, or 'paying attention' to each other in a group, also translates to humans who spend most time communicating and interacting with people we call friends. Or, to flip it around, our friendship group consists of the people with whom we currently spend most time interacting and socializing. Human acquaintances can also be mapped in both directions from the magic 150—with inner circles of closer and closer relationships, a core circle of five in the centre, and then outer circles of 500 acquaintances, and 1,500 people for whom we can put a name to a face (see Figure 44).

Our expensive brain

Being able to control, manipulate, and/or influence groups of individuals must have provided high rewards because there are also high costs associated with having a large brain. The first is the problem and danger of giving birth to a baby with a large head. This is called the obstetric dilemma. Chimpanzees give birth relatively easily as the size of the head is significantly smaller than the birth canal (Figure 45). Moreover, the baby comes out face forward so that it is relatively easy for the mother to pick up the baby and lift it directly on to the breast for suckling. With A. afarensis, the changes in the hips to allow bipedal locomotion, and the slightly larger head means the baby's head comes sideways through the birth canal. Again, it is relatively easy for the mother to reach down, pick up the baby, and place it on her breast (Figure 45). In H. sapiens, the head of the baby relative to the birth canal is much larger, and so the baby head has to twist twice to get through the hips. Like A. afarensis, a

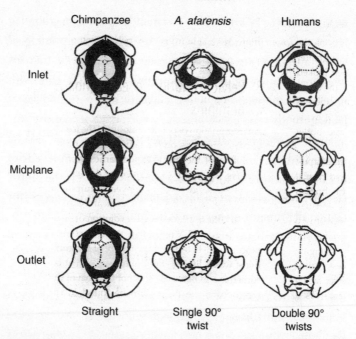

FIGURE 45 A 'midwife's' view of the birth canal in a chimpanzee, A. *afarensis*, and a modern human. Note that human babies undergo two 90° twists and usually are born facing their mother's bottom.

modern human baby twists sideways to enter the birth canal, but halfway through it has to twist again to get past the hip bones. This means that the majority of H. *sapiens* babies are born facing the mother's bottom. This makes it very awkward for the mother to reach down to pick up the baby and twist it round so it faces her breast for suckling without getting it tangled with the umbilical cord. Palaeo-anatomists think that H. *erectus* was probably the first hominin that had the double twist birth, which must have evolved due to a significant increase in brain size. This does mean that childbirth in later *Homo* species was more dangerous to the mother and infant than for apes and earlier hominins. This is due to the tightness of fit of the baby's head and the double twist, which adds complications such as the umbilical cord getting wrapped around the infant's neck. Even today with modern hospitals, childbirth can occasionally be fatal for mother and/or child.

So there are two fundamental questions we have to ask to understand why the obstetric dilemma occurred. First, why cannot the heads of babies be smaller? At the moment about 30 per cent of adult brain size is achieved, on average, at birth. It seems that, in all primates, 30 per cent is the minimum required to produce a viable primate infant. This is supported by modern medical data, which suggests the prevalence of special educational needs in children rises sharply in babies born prematurely. Hence, producing small-brained babies seems a non-viable option. Also, it is not just the head that can cause problems in childbirth but also the shoulders and relatively large body weight of the baby, both of which can lead to maternal tearing and haemorrhaging and muscular or spinal damage of the offspring. However, birth weight is the single biggest predictor of survival in early life; so any reduction in non-brain body weight will adversely affect the survival chances of the infant.

The second question is: why can't the mother's pelvis be larger? The pelvis is limited by the need to allow bipedal movement, but there is variation in human pelvis size both front to back and side to side; so the mother's pelvis in some cases could be larger. These modern variations seem to be tied to local environmental conditions and mothers' access to nutrition as they are growing up. For example, Jonathan Wells at University College London points out that African and Asian women have a narrower pelvis than

European women in response to hotter, more humid environments, but correspondingly have slightly shorter gestation periods. Within the African population, short mothers have the highest risk of obstructed labour, suggesting their pelvis could have been larger if they had had the required nutrition for full growth during their own development. There is now evidence that the balancing act between ensuring that the baby is large enough to survive early life and avoiding its becoming too big to fit through the pelvis are negotiated through metabolic signals between the mother and the foetus. Interestingly, Jonathan Wells notes that the current large burden of maternal mortality around the world is due partly to the underdevelopment of many mothers' pelvises because of undernutrition early in life, and partly to increased foetal birth weight related to the emerging obesity epidemic resulting from rapidly changing diets—both of which are disturbing the balance achieved by the obstetric dilemma.

During the period of rapid evolution, when brain size was expanding significantly and the pelvis had to change shape to accommodate it, the obstetric dilemma would have been a continual process and must have resulted in a large increase in maternal deaths. So there must have been huge rewards for being smart and having a bigger head for the selection pressure to continue. This has been called the *expensive brain framework* (Figure 46), which tries to understand the advantages of having enhanced cognitive ability. One of the most interesting things about the complicated childbirth in later *Homo* is that mothers would have required help, called allomaternal care. So individual females who were more socially adept would get more help, and they and their infants were more likely to survive—another positive feedback loop driving the evolution of larger brains as a means of having greater

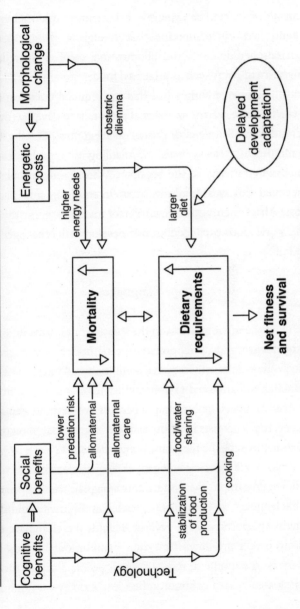

FIGURE 46 Expensive brain hypothesis.

social influence. The expensive brain framework tries to understand the positive effects, such as food production and sharing, predation reduction, and allomaternal care, compared to the negative impacts such as increased food requirement and mother mortality. Underlying all of this is the social brain hypothesis, which can be seen as an internal arms race to develop the higher cognitive skills to enable greater social control. Clearly, with the emergence of *Homo erectus*, *H. heidelbergensis*, and *H. sapiens*, the positives outweighed the negatives. We can only speculate as to whether this was driven by stressful environmental change in East Africa, increased competition for resources within the species as populations expanded, or competition with other species—or all of these.

Language

One of the missing pieces in the story of human evolution is the appearance and development of language, as the social brain hypothesis implicitly requires complex social interaction—or, as Robin Dunbar would say, the ability to gossip. I have not mentioned language before simply because we have no idea when it appeared. Scientists have strong views on this, based on their different theoretical backgrounds and their views on how modern human babies develop language skills. The views can be split into two main camps. The first are continuity theories, which suggest that language now exhibits so much complexity it could not simply have appeared from nothing. It must therefore have evolved from earlier pre-linguistic systems among our hominin ancestors. At the moment, most linguistic scholars accept the underlying idea behind continuity theories, but they vary in how they envisage language development. Steven Pinker, experimental psychologist and cognitive scientist at Harvard University, is among those who see language as mostly innate, having evolved in a gradual way.

Other scientists support discontinuity theories, according to which language is a unique trait that cannot compare with anything found among non-humans, and must therefore have appeared suddenly with *Homo sapiens*. Noam Chomsky, linguist and cognitive scientist, formerly at MIT, is a prominent proponent of discontinuity theory. He argues that a single chance mutation occurred in one individual around 100,000 years ago, instantaneously installing the faculty of language in a 'perfect' or 'near-perfect' form. This view is supported by Ian Tattersall of the American Museum of Natural History, who argues that evidence of symbolic reasoning does not appear in the archaeological record until after 100,000 years ago, and these early cultural artefacts are evidence for the spontaneous invention of language.

I would, however, counter this by pointing out that there are examples of symbolic reasoning prior to *H. sapiens*. For example, Josephine Joordens and colleagues at Leiden University published, in *Nature*, an engraved freshwater mollusc pseudodon shell found at a site known as Trinil on the island of Java. It was almost certainly etched by *Homo erectus* around 500,000 years ago, showing the ability for abstract design and thus symbolic activity. While in Germany, beautifully preserved wooden spears have been found that were crafted and used by *H. heidelbergensis* 400,000 years ago. It is difficult to see how these weapons could have been crafted without teaching, and that requires sophisticated communication techniques. My own view is that it is difficult to see how expansion of the brain and our ancestors' cognitive abilities could have occurred without some form of language. One lesson we have learnt in studying human evolution is that the archaeological record is incomplete, and many stages of development in fact occured much earlier than our records show. Stone tools are now associated with *A. afarensis*, so it would not surprise palaeoanthropologists if evidence were found for symbolic behaviour when *H. erectus* first appeared, which would then mean that language was central to the development of the genus *Homo*.

Sexual behaviour and human evolution

Another facet of human behaviour, which may have co-evolved with our large brain, is our mating strategies. Primates exhibit all sorts of mating behaviour, including polygynous, multimale–multifemale, and monogamous. One control on which behaviour is selected is the degree of size difference between males and females. The greater this sexual dimorphism, the more likely the mating strategy is either polygynous or multimale–multifemale. This can be shown by observing our closest living relatives—chimpanzees and gorillas. According to calculations using the rate of DNA mutations, it has been calculated that our common ancestor with chimpanzees lived about 8 to 5 million years ago (see Box 4). There is a large difference in size between male and female chimpanzees, and they have a multimale–multifemale mating system. Essentially, male chimpanzees have sex with any female, all the time, and with any excuse. Hence, a female may contain sperm from multiple partners at any one time. So, despite close genetic ties to humans, chimpanzees have adaptations for direct sperm competition. For example, their testicles are huge, to facilitate massive production of sperm multiple times a

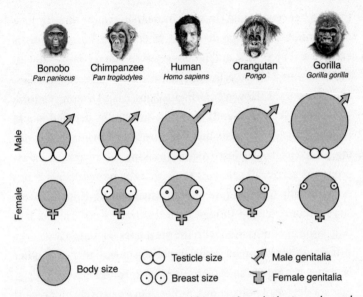

| Bonobo | Chimpanzee | Human | Orangutan | Gorilla |
| *Pan paniscus* | *Pan troglodytes* | *Homo sapiens* | *Pongo* | *Gorilla gorilla* |

Body size Testicle size Male genitalia
Breast size Female genitalia

FIGURE 47 Comparison of great ape and human relative body size and sexual organs.

day. A chimpanzee's testes weigh more than a third of its brain (Figure 47). In contrast, human testes are only about 3 per cent of the weight of an adult's brain.

There is also a large difference in size between male and female gorillas, but they have a polygynous or harem-style mating system—that is, many females to a single male. Gorillas, like humans, therefore have relatively small testes. This suggests that sperm competition is not the primary evolutionary driving force and thus there have been no special adaptations of the male reproductive system to facilitate this. This is similar to what is found in modern humans, whose reproductive glands or testes are of a very modest size (Figure 47); their sperm count reduces by more than 80 per cent if men ejaculate more than two or three times a day. Some experts have discussed whether the human penis is exceptional in its size, and it does seem so when compared with those of our closest primate relatives—chimpanzees, gorillas, and orangutans. However, primatologist Alan Dixson (Victoria University, New Zealand) in his wonderfully detailed book, *Primate Sexuality*, suggests that if we look at all primates, including monkeys, this is just wishful thinking. Comparative measurements across all primates show that the human penis is not exceptionally long, and its only unusual feature is its relatively large circumference—though even this trait is not unique. It is only large in comparison with the great apes (Figure 47).

If we compare human sexual dimorphism with that of all other primates it suggests an evolutionary background involving a significant degree of polygynous, rather than exclusively monogamous, mating. This is supported by anthropological data studying recent human populations, which shows that most populations engage in polygynous marriage. Anthropologists Clellan Ford and Frank Beach, in their book *Patterns of Sexual Behavior* first published in 1951, suggested that 84 per cent of the 185 human cultures they had data on engaged in polygyny. However, it is important to note that it is usually only high-status or wealthy men who are likely to have polygynous marriages; the majority of people engage in monogamy, even though they are in a polygynous society. In comparison, monogamy and serial monogamy is the main system used by hunter-gather bands around the world, which might provide a better analogy to our ancestors.

The problem is that the fossil record leaves no trace of our ancestors' sexual behaviour. However, comparing hominin sexual dimorphism the fossil evidence is clear that the *Australopithecus* and *Parathropus* genera had a much larger dimorphism than the *Homo* genus. This suggests that early hominins may have had a more polygynous mating system than later hominins. It is also apparent that the sexual dimorphism of *H. erectus* was very similar to that of modern humans, implying that human mating systems may have evolved early on, at the same time as there was a significant increase in brain size. Kit Opie and colleagues at University College London analysed the behaviour of 230 primate species to understand why a large proportion of human mating is monogamous or, at the very least, serial-monogamous. They argue that social monogamy or pair-living is very uncommon in mammals—less than 3 per cent—compared to birds, for which it is over 90 per cent, though in primates a quarter of the known species do engage in some form of social monogamy. Three main theories have been put forward for social monogamy: parental care, mate guarding, and infanticide risk. For males, it is advantageous to seek as many opportunities to reproduce as possible, so something has to keep males hanging around. Kit Opie's analysis suggests that because human infants are so vulnerable for so long, infanticide by other males is a big risk. For example, it is well known that when a younger male lion finally displaces the alpha male lion in charge of a pride, they will systematically kill all cubs of the previous male to ensure that all new offspring are their own. Hence, to ensure that children are able to reach maturity the male is likely to stay to protect them, both socially and physically, if needed. In contrast, zoologists Dieter Lukas and Tim Clutton-Brock at Cambridge University argue that it is less to do with infanticide but rather linked to males being unable to defend access to multiple females.

If we view the evolution of monogamy or serial monogamy mating systems in humans through the lens of the social brain hypothesis, it is clear that it takes a huge amount of social effort to maintain and protect more than one mate at a time. It is only in more advanced societies, where males have access to wealth and power, that they can protect multiple females, usually by paying other men to protect them. So monogamy seems to be an adaptation to protect one's mate and children from other males. This is re-enforced by the high social cost and stress of attempting to do this for multiple partners, and it has been supported in complex human societies by cultural norms.

The *Homo sapiens* problem

There are two problems with *Homo sapiens*. The first is that we have very little idea why *H. sapiens* emerged between 300,000 and 200,000 years ago. There is evidence that the climate of East Africa, after being very dry and stable for 400,000 years, became more variable, and this may have driven the small evolutionary changes from *H. heidelbergensis* to *H. sapiens*. But we may also have to reassess our assumption of a single evolutionary occurrence of *H. sapiens*. The new idea emerging is that evolution of *H. sapiens* populations in Africa was multi-regional. This is because over the last 300,000 years, the fossils show a complex mix of archaic and modern features in different places and at different times. Physically and culturally there is a continental-wide trend towards the modern human form, but this varies spatially and temporally—and there seems to be no consistency until about 60,000 years ago. Again we return to idea that human evolution is a deeply interconnected bush—not a simple tree. The second problem is, for me, the more intriguing: that after *H. sapiens* evolved in Africa and spread out into Europe and Asia they seemed to do nothing special for at least the first 150,000 years. Over these 150,000 years there are increasing records of symbolic behaviour, starting with microliths, shell engravings, ochre, and shell beads. But it is not until about 50,000 years ago that consistent signs of creative thinking emerge—beautiful cave paintings in Spain, France, and Indonesia, beautifully carved Venus figurines in Germany, the Czech Republic, Austria, France, and Siberia, and shell beads in North Africa and Europe. Around the same time, modern humans appear that were more slender than their earlier ancestors, and had less hair and smaller, less robust skulls—they looked basically like us. But these changes weren't just cosmetic, as the appearance of these smaller, more fine-boned humans was accompanied by a revolutionary development of cumulative culture that led to the birth of agriculture and eventually to human dominance of the planet.

There are, however, some tantalizing hints of what may have happened from several very recent studies. The first is an analysis of the fossilized skulls of our ancestors during this transitional period, carried out by a team led by Robert Cieri at the University of Utah and published in the journal *Current Anthropology*. Cieri and colleagues found that the brow ridge (the bony bit above the eye sockets) became significantly less prominent and male facial shape became more similar to that of females. They referred to this as craniofacial feminization, meaning that, as *H. sapiens* slimmed down, their skulls became flatter and more 'feminine' in shape. They think this must have been due to lower levels of testosterone, as there is a strong relationship between levels of this hormone and long faces with extended brow ridges, which we may perceive today as very 'masculine' features (Figure 48). There is a second line of evidence that comes from studying the relative finger lengths of our ancestors. There seems to be a strong correlation between the ratio of the length

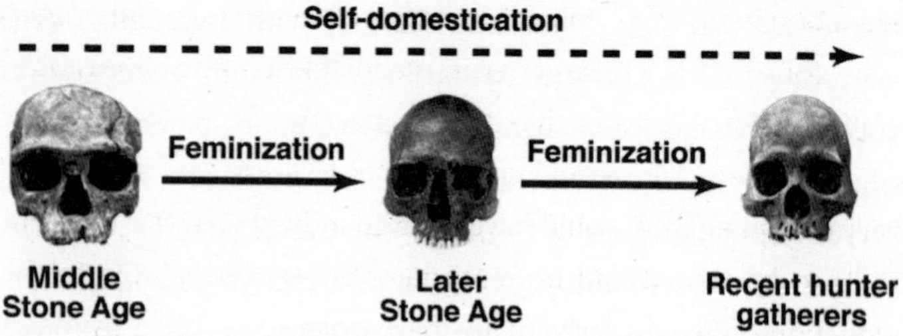

Self-domestication

FIGURE 48 Comparison of *Homo sapiens* skull morphology showing feminization that has occurred since the Middle Stone Age.

of the second (index) and fourth (ring) fingers to aggression, promiscuity, and competitiveness in humans. The primary reason for this relationship is because the digit ratio seems to reflect prenatal testosterone levels. A hand with a shorter index finger than the ring finger suggests higher testosterone levels. Emma Nelson from Liverpool University and Susanne Shultz, now at Manchester University, have found this relationship also holds across the great apes and New World monkeys. Human digit ratios are intermediate between pair-bonded and the more promiscuous great apes, supporting the idea that monogamy is partly socially constructed to protect mates and children. Moreover, there is a clear drop in digit ratio at the same time as craniofacial feminization occurs, suggesting that there is a significant drop in prenatal testosterone levels during this transitional period.

People with lower levels of testosterone are less likely to be reactively or spontaneously violent, and therefore this would have greatly enhanced social tolerance. This has a huge knock-on effect. As seen among humans today, we live in extremely high-density populations with an incredible amount of social tolerance. So a reduction in reactive violence must have been an essential prerequisite for our living in larger groups and developing a more cooperative culture. The idea that humans became more feminine, less aggressive, and thus could cooperate in large groups is certainly very intriguing as it would have allowed individuals with different skills to be valued and be reproductively successful due to the reduction of, particularly, male–male violence. In most primates the physically strongest male tends to dominate, but in early humans the smartest or the most creative people may have come to the forefront.

Self-domestication of humans

The question remains: how did we become more feminine, less violent, and more creative? A paper in the journal *Animal Behaviour* by Brian Hare at Duke University and his colleagues may throw some light on this. They compared chimpanzees (*Pan troglodytes*) and bonobos (*Pan paniscus*) in West Africa—two closely related species living in very similar environmental conditions either side of the Congo River. One key distinction between the two species is their sexual dimorphism. Male chimps are significantly larger than females, whereas the difference in bonobos is much smaller. This difference is driven by different levels of testosterone. Size is just one manifestation of deeper differences that show up in how the animals interact with one another. Chimpanzees, particularly males, are very aggressive; but violence within or between groups is almost non-existent among bonobos. As both these species have a common ancestor there must have been strong selection going on to feminize the bonobos. Hare and his colleagues suggest a process of self-domestication, whereby violent individuals

are punished and prevented from reproducing. The traits exhibited by bonobos are very similar to the changes observed in species that humans have domesticated, such as dogs, cows, guinea
pigs, and foxes. They postulate that the reason bonobos were able
to feminize and chimpanzees were not is because on the eastern
side of the Congo, where the chimps live, they are in direct competition with gorillas, whereas the bonobos on the western side
have no competition. Harvard professor Richard Wrangham, a
co-author of the Hare paper, suggested that the same process may
have happened to early humans.

Equality improves networking

This feminization through self-domestication may not only have
made groups of humans more peaceful, but may have also produced a more sexually equal society. A recent study in *Science* by colleagues of mine at University College London showed that in
hunter-gatherer groups in the Congo and the Philippines decisions
about where to live and with whom were made equally by both gen-
ders. Despite living in small communities, this resulted in hunter-
gatherers living with a large number of individuals with whom they
had no kinship ties. The authors argue this may have proved an evolutionary advantage for early human societies, as it would have fostered wider-ranging social networks, closer cooperation between
unrelated individuals, a wider choice of mates, and reduced chances
of inbreeding. The frequent movement and interaction between
groups also fostered the sharing of innovations, which may have
helped the spread of culture. As Andrea Migliano, the leader of the
study, points out, 'sex equality suggests a scenario where unique
human traits, such as cooperation with unrelated individuals, could

have emerged in our evolutionary past'. It may have only been with the rise of agriculture that an imbalance between the sexes re-emerged, as individual men were suddenly able to concentrate enough resources to maintain several wives and many children. Indeed, the Robert Cieri-led study does show slightly more mascu-line facial shapes emerging in recent agriculturalists relative to early humans and recent human foragers.

So at the moment we have some tentative hints of what may have happened between 50,000 and 10,000 years ago. Humans may have undergone self-domestication and, over many gener-ations, weeded out those individuals who were unable to control their reactive violence. This is not as far-fetched as it sounds—studies of the Gebusi tribe in Papua New Guinea by Bruce Knauft showed significant levels of male mortality due to the tribe decid-ing that if an individual's behaviour was intolerable, then for the good of the tribe they must be killed. So human proactive violence—that is, thought-out, discussed, and planned violence—is used to curb, control, and cull reactively violent individuals. This process, combined with female mating choices which, instead of selecting just for larger cognitive ability selected males which dominated the group socially and not violently, over thou-sands of years, could have resulted in males with lower testoster-one and more feminine features, leading to a much more gender-equal society and the start of our cumulative culture.

Dispersal of modern humans

The next stage in the *H. sapiens'* takeover of the world was their dispersal out of Africa into all the major continents, except Antarctica. The evidence suggests that *H. sapiens* first evolved in

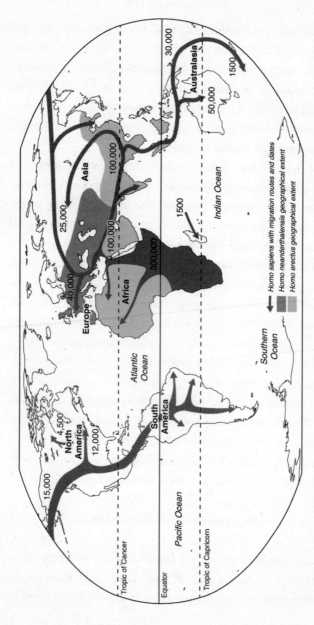

FIGURE 49 Dispersal of *H. erectus*, *H. Neanderthalensis*, and *H. sapiens*. Dates are in years before present and refer to the global dispersal of *H. sapiens*.

Africa between 300,000 and 200,000 years ago; they then dispersed out of Africa into the Middle East about 120,000 years ago. New evidence from fossilized teeth in China suggests that modern humans may have made it all the way to China from between 100,000 and 80,000 years ago. Between 60,000 and 40,000 years ago, humans made it into Australia. Then, about 40,000 years ago, humans made it into Europe and started to compete for resources with *Homo neanderthalensis*. Genetic studies have also shown that this European expansion went east through what is now Russia, and formed the base population which migrated into Mongolia and Korea, and spread across the Bering Sea into North America, Central America, and then South America (Figure 49). When *H. sapiens* encountered other hominin species such as Neanderthals and Denisovan there seem to have been multiple periods of interbreeding which has led to a genetic legacy of those species within our own genome (see Box 4).

Neanderthals

Current evidence suggests that Neanderthals and humans shared a last common ancestor sometime between 765,000 and 550,000 years ago. We speculate that this ancestor was *H. heidelbergensis*. The story gets a little bit more complicated as there are also Eastern Neanderthals, called Denisovans, whose genetics seem to indicate a close affinity to European Neanderthals (Figure 50). Evidence for *H. denisovans* has been found in Siberia, Asia, and Melanesia. One major difference between early African humans and Neanderthals was their diet. This is because the cold, sparse winters in Eurasia would have restricted plant availability, forcing Neanderthals to rely more heavily on meat—indeed, chemical

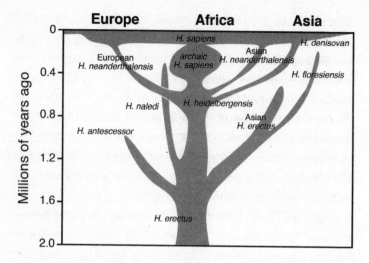

FIGURE 50 Relationship between *Homo erectus* and *Homo sapiens* based on current fossil and genetic evidence.

analysis of their bones shows that it dominated their diet. It seems Neanderthals were specialized seasonal hunters, and there is evidence for them killing reindeer in winter and red deer in summer. Neanderthal archaeological sites contain sharp wooden spears and large numbers of bones of big game animals, showing evidence of hunting and butchery. Neanderthals were, however, adaptable, and at the Gibraltar coastal site there is evidence they exploited marine resources such as fish, molluscs, seals, and even dolphins. They did also eat plants, and remains of starch grains have been found in excavated Neanderthal molar teeth.

Neanderthals also had a sophisticated stone tool industry. This differs from earlier 'core tool' traditions, such as the Acheulean tradition of *H. erectus*. Acheulean tools are created by removing flakes from the surface to 'reveal' a tool. In contrast, the Neanderthal Mousterian stone tool industry is characterized by flake tools that were detached from a prepared stone core. This innovative technique allowed multiple tools to be fashioned from a single suitable stone. Neanderthals used tools for activities such as hunting and sewing. There is evidence for left–right arm asymmetry, which was originally assumed to have developed from the hunting technique of thrusting spears instead of throwing them. This is in contrast to other hominins, such as *Homo erectus, H. heidelbergensis,* and *H. sapiens,* that had shoulder adaptations to allow them to throw weapons, and in the case of the later two species there is archaeological evidence for thrown spears. The Neanderthals' style of hunting at close quarters has been used to explain why Neanderthal bones have such a high frequency of fractures, as these injuries are similar to those among professional rodeo riders who regularly interact with large, dangerous animals. Colin Shaw from Cambridge University, however, has suggested a less glamorous reason for asymmetry between the arms. He suggests that Neanderthals may have spent hours scraping animal hides for clothing production, using stone tools. This would have been an arduous and repetitive task, but essential in the cold conditions in which they lived. The asymmetry in the arms of Neanderthals has no direct analogy in modern humans but some sports people, such as cricketers and tennis players, do have very pronounced one-sided development.

One of the reasons Neanderthals are so interesting to palaeoanthropologists is because they are the first species to show evidence of wearing clothes and of consistent symbolic behaviour. Scrapers and stone and bone awls have been found, along with animal bones, at Neanderthal sites. It is suggested the Neanderthals would have used a scraper to first clean the animal hide, then an awl to poke holes in it, and strips of animal tissue to lace together a loose-fitting garment. It is only with modern humans that we find evidence of the bone sewing needles needed to make tighter fitting clothing. Neanderthals also controlled fire and lived in shelters. There is evidence that Neanderthals deliberately buried their dead and occasionally even marked their graves with offerings, such as flowers. This may also explain why we have such a rich history of Neanderthals, as the burials greatly enhance the chances of preservation. They also made symbolic and ornamental objects such as grooved and perforated animal teeth, mammoth-ivory rings, decorated bone tools, and used colourants. No other species had ever practised this sophisticated and symbolic behaviour.

Neanderthals and modern *Homo sapiens* may have inhabited similar geographic areas in Eurasia for nearly 50,000 years. They may have had little direct interaction for most of this time as recent genetic evidence suggests that the Neanderthal population was very small compared with *H. sapiens* over the 400,000 years of their existence. There is, however, clear genetic evidence that the two species did interact at some stage. Scientists have recently sequenced Neanderthal mitochondrial and nuclear genomes. When compared with modern humans, it seems that many non-African people have between 2 and 4 per cent Neanderthal ancestry, meaning that Neanderthals and early humans must have interbred. It is also clear that Europeans and Asians have inherited Neanderthal genetic material, while some Asians have also inherited Denisovan genetic material (Box 4). This admixture, though small, may have provided adaptive advantages to non-African humans. Just a few thousand years after modern humans moved into Europe, Neanderthal numbers dwindled to the point of extinction. All traces of Neanderthals disappeared about 40,000 years ago. There

Box 4: Genetic revolution

Since the 1960s, the field of evolutionary genetics has arguably made the greatest impact of any discipline upon human evolutionary studies. Advances in technology, such as the development of the polymerase chain reaction, which can amplify small quantities of DNA, have been instrumental in this revolution, allowing, for example, the analysis of ancient DNA taken from Neanderthal fossils. Allan Wilson, Emile Zuckerkandl, and Linus Pauling pioneered the use of molecular approaches, which examined evolution at the scale of DNA and proteins, showing that relationships among living and extinct primates can be inferred from genetics as well as fossils. DNA can also be used as a 'molecular clock', involving comparison of the amount of genetic difference (mutations) between living organisms. Since mutations accumulate at approximately predictable rates over time, they can be used to estimate how long ago two living species shared a common ancestor. The molecular clock cannot assign concrete dates and must be calibrated against independent evidence, such as the fossil record. Nevertheless, bringing together the transdisciplinary evidence, we now have a robust understanding of the relationships between humans and apes. For example, we know that humans and chimpanzees split from their common ancestor approximately 8 to 4 million years ago, and that the genetic difference between humans and chimpanzees is miniscule, at about 1.2 per cent.

Genetics has also confirmed that living humans have a limited genetic diversity, indicating that there may have been a series of population bottlenecks. These temporary, drastic reductions in population size and therefore genetic variability may have been caused by rapid climate changes, earthquakes, or even disease. Studies of genetic variation reveal that the greatest diversity can be found in African populations. This, combined with evidence from histories following the female (mitochondrial DNA) and male (Y chromosome) lines, confirms

continued >

an African origin for our species and suggests that our direct ancestors migrated out of Africa between 70,000 to 40,000 years ago.

Genetics has also contributed to our understanding of the complex relationship and interbreeding between *H. neanderthalensis*, Denisovans, and *H. sapiens*, pushing the boundary of DNA analysis to more than 400,000 years. It seems that there may have been at least five inbreeding events in the last 0.5 million years (Figure 51). Genetics is demonstrating that human evolution, rather than being a simple linear evolution between species, is much more a network of slightly different species interbreeding and exchanging DNA. We need to see human evolution not as an evolution tree but more of a tangled bush with

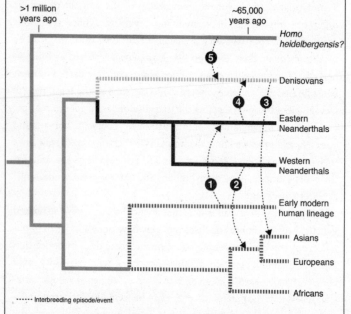

FIGURE 51 A history of interbreeding between early modern humans, Denisovans, and Neanderthals.

continued >

many interconnections. We can speculate that the interbreeding that occurred in the most recent hominins may have also occurred earlier in our history, explaining why there were so many very morphologically similar hominins coexisting in East Africa in the past. Genetics is also making contributions towards revealing the evolution of human cognition, for example language capacity (e.g. the FOXP2 gene) and identifying human-specific neurological structures (e.g. the HARs, SRGAP2, GPRIN2, GTF2I, and HYDIN2 genes).

is an endless debate as to whether *H. sapiens* simply outcompeted Neanderthals, or whether there was proactive culling, or both.

By 40,000 years ago we think that *H. sapiens* was the last hominin species having outcompeted, interbred, and even killed off the other hominin species. We humans are the only ultrasocial creature on the planet. We co-habit in cities of tens of millions of people and violence between individuals is extremely rare—even though this is not the picture painted by the media. So when we study human evolution we need to understand the cause of the development of our extremely large, flexible and complex 'social brain'. Of course, we can see many advantages in having a large brain. First, it allows humans to exist in a group size of about 150, which builds resilience to environmental changes by increasing and diversifying food production and sharing. Humans have no natural weapons, but living in large groups and having tools allowed humans to become the apex predator, hunting animals as large as mammoths. Larger groups also offer more protection from other predators. Second, it allows specialization of skills such as support for childbirth, tool-making, and hunting.

There is new genetic evidence by Aida Gomez-Robles and colleagues at The George Washington University that suggests the modern human brain is indeed extremely flexible. They showed that the human brain is more plastic and is modelled more by the surrounding environment than that of chimpanzees. The anatomy of the chimpanzee brain is strongly controlled by their genes, whereas the human brain is extensively shaped by the environment, no matter what the genetics. This means the human brain is pre-programmed to be extremely flexible; its cerebral organization is adjusted by the environment into which it has been born and raised. So each new generation's brain structure could adapt to new environmental challenges without the need to evolve physically. This may also explain why we all complain that we do not understand the next generation, as they have different brain structures from us because they have grown up within a different physical and social environment.

Having a large brain does not mean that culture and society were inevitable—far from it, as there seem to be other requirements before human culture could start to build up. These may have involved a reduction in reactive violence, and populations of humans simply reaching a size at which inventions and new ideas were not lost, and real accumulation of knowledge could occur.

But once humans had a cumulative culture, then things started to change—first with the agricultural revolution and later with the industrial revolution.

9

The Future of Humanity

Human evolution has clearly been heavily influenced by tectonics and celestial mechanics through local environmental change. Through the middle to the end of the last ice age, *H. sapiens* dispersed into Australasia, Europe, and the Americas. Once *H. sapiens* acquired cumulative culture, in the middle of the last ice age, they started to have a larger influence on their local environment. At the end of the last ice age, 11,000 years ago, agriculture first appeared in South West Asia, South America, and north China. Independently, it then appeared 7,000–6,000 years ago in south China and Central America, and 5,000–4,000 years ago in the savannah regions of Africa, India, South East Asia, and North America. There is evidence that this early agriculture influenced atmospheric greenhouse gases, with the release of carbon dioxide from deforestation and the release of methane through wet rice cultivation (Figure 52). *H. sapiens*' influence on the local and global environment has accelerated through the industrial revolution to the present day. Our influence is now so great that scientists suggest we are a major force of geology, as powerful as

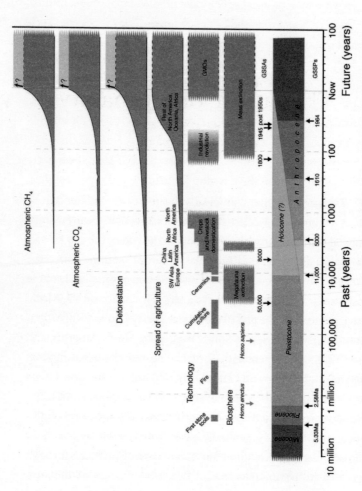

FIGURE 52 Mass extinctions, land-use changes, and rising atmospheric greenhouse gas levels over the past 50,000 years.

tectonics or a meteorite impact. To acknowledge our dominant role in the global environment, scientists have suggested we have entered a new geological epoch called the Anthropocene. In this chapter I will examine the evidence for the Anthropocene, and what it means for human evolution.

Evidence for the Anthropocene

Are we really that important on planet Earth? Has, as Desmond Morris called us, a *naked ape* really become the dominant geological power on the planet? Let's look at the evidence. If we start with the atmosphere there is clear evidence that human actions have released 555 petagrams of carbon (where 1 Pg = 10^{15} g = 1 billion metric tonnes) into the atmosphere since 1750, increasing atmospheric CO_2 to a level not seen for at least 800,000 years, and possibly several million years (Figure 52). There is even evidence that these extra greenhouse gases have delayed the Earth's next ice age. The released carbon dioxide has also increased ocean water acidity at a rate probably not exceeded in the past 300 million years. Despite the focus on greenhouse gases due to concerns about climate change, humans have also profoundly affected other parts of the Earth system—for example, the nitrogen cycle. In the early twentieth century, the invention of the Haber–Bosch process allowed us to convert atmospheric nitrogen to ammonia for use as fertilizer, revolutionizing agriculture and helping to feed billions of people. But it has also altered the global nitrogen cycle so fundamentally that the nearest suggested geological comparisons are events about 2.5 billion years ago.

There is clear evidence that anthropogenic greenhouse gases are changing our climate. These changes include nearly a 1°C increase in average global temperatures and sea-level rise of over 20 cm in the past 100 years. There is also evidence for significant shifts in the seasonality and intensities of precipitation, changing weather patterns, and the significant retreat of Arctic sea ice and nearly all continental glaciers. It is estimated that Greenland is losing over 200 gigatonnes of ice per year, a six-fold increase since the early 1990s; Antarctica is losing about 150 gigatonnes of ice per year, a five-fold increase since the early 1990s—and most of this loss is from the northern Antarctic Peninsula and the Amundsen Sea sector of West Antarctica.

Human drivers of evolution

Human actions are also affecting non-human life. Over recent decades, global net primary productivity—the amount of new plant growth each year—appears to have been relatively constant. However, humans appropriate between 25 per cent and 38 per cent of this net primary productivity, which of course reduces the amount available for millions of other species on Earth. In a paper in *Nature*, Thomas Crowther and colleagues at Yale University estimated that before the industrial revolution there were 6 trillion trees on Earth, but now there are only 3 trillion trees left. This deforestation and land-use conversion to produce food, fuel, fibre, and fodder, combined with targeted hunting and harvesting, has resulted in species extinctions some 100 to 1,000 times higher than background rates, and probably constitutes the beginning of the sixth mass extinction in Earth's history. Species removals are non-random, with a disproportionate removal of animals with larger body sizes from both the land and the oceans. This Megafauna Extinction began between 50,000 and 10,000 years ago, depending on when modern humans arrived on the continent. Overall, during the Megafauna Extinction about half of all large-bodied mammals worldwide, equivalent to 4 per cent of all mammal species, were lost. The losses were not evenly distributed, as Africa lost 18 per cent, Eurasia lost 36 per cent, North America lost 72 per cent, South America lost 83 per cent, and Australia lost 88 per cent of their large-bodied mammals. What may surprise you is the massive shift from wild mammals to humans and domestic mammals since this extinction began. According to Vaclav Smil of the University of Manitoba, the biomass of land mammals on Earth today is made up of 30 per cent humans and 67 per cent domesticated animals; wild terrestrial mammals make up just 3 per cent (Figure 53).

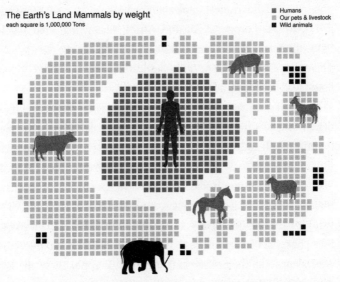

FIGURE 53 The Earth's land mammals by weight.

Organisms have also been transported around the world, including crops, domesticated animals, and pathogens on land. Similarly, boats have transferred organisms among once-disconnected oceans (Figure 54). Such movement has led to a small number of extraordinarily common species, new hybrid species, and a global homogenization of Earth's biota. Just one example of this is that nearly all the earthworms in North America are European in origin because when they were inadvertently brought over by the early colonists in the sixteenth century they outcompeted the local worms, and took over.

By far the biggest change was started in the sixteenth century, when species from the Old World (Europe and Asia) were exchanged with those in the New World (the Americas). This is known as the Colombian Exchange. One result of the exchange was the globalization of human foodstuffs. The New World crops maize/corn and potatoes, and the tropical staple manioc/cassava, were subsequently grown across Europe, Asia, and Africa. Meanwhile, Old World crops, such as sugarcane and wheat, were planted in the New World. There was also movement of domesticated animals such as horses, cows, goats, and pigs, all to the Americas. Human commensals, such as the black rat, also made it to the Americas. There were other accidental transfers from Europe too, such as smallpox, measles, and typhus, which ended up killing over 50 million people in South and Central America in the sixteenth century. These exchanges of plants and animals continue today, and invasive species have become a major concern on all continents. A study in *Nature* by ecologist Mark van Kleunen and colleagues at the University of Konstanz suggests that 4 per cent of plant species have been relocated around the globe, equivalent to all the native plant species in Europe. These changes are unique, since the supercontinent Pangaea separated about

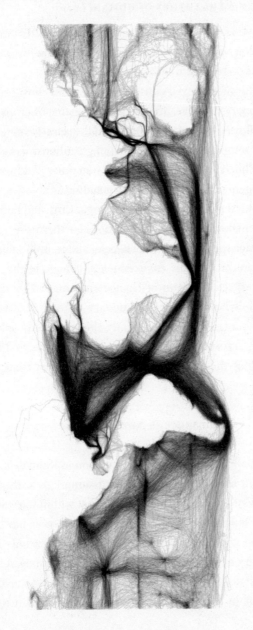

FIGURE 54 Shipping routes plotted from eighteenth- and nineteenth-century logs.

200 million years ago, but Simon Lewis and myself at University College London would argue that such trans-oceanic exchanges probably have no geological analogue.

Human actions may well constitute the Earth's most important evolutionary pressure. The intense harvesting of crops, the movement of species to new habitats, and the selective pressure of higher air temperatures resulting from greenhouse gas emissions are all likely to alter evolutionary outcomes. Add to this the development of diverse products, including antibiotics, pesticides, and novel genetically engineered organisms, and humans are now the major influence on evolution. Furthermore, given that the average species survives between 1 and 10 million years, the rates of anthropogenic environmental change in the near future may exceed the rates of change encountered by many species in the whole of their evolutionary history. So I would suggest that human activity has clearly altered the lithosphere, hydrosphere, cryosphere, atmosphere, and biosphere, and thus we are changing the environmental and evolutionary trajectory of the Earth.

Anthropocene paradigm shift

The physicist and philosopher of science Thomas Kuhn defined a paradigm shift as a change in the basic assumptions within the ruling theory of science. I would contend that within science the evidence for humans being a major geological power has been accepted, and the paradigm shift has already occurred. That human activity has altered Earth fundamentally is rarely, if ever, questioned in scientific publications. But I do want to be clear that this shift in the scientific paradigm, though it is through complex social processes, is *a better*, not just a different, understanding of the world. This is because a common misinterpretation of paradigms is the belief that the discovery of paradigm shifts and the dynamic nature of science is a case for relativism— i.e. that science only has subjective value according to differences in perception, consideration, or beliefs. Thomas Kuhn, and most scientists, would vehemently deny this as the rational assessment of the weight of scientific evidence means the new paradigm, if evidence-based, is always superior to the previous theory.

What is now being argued about is exactly when the Anthropocene Epoch began, so it can be ratified and added to the official geological timescale. Now, as I am sure you would expect, the arguments about when humans became the dominant geological power on Earth are robust, to say the least. The three main contenders are the influence of early agriculturists from 5,000 years onwards, the Colombian Exchange, with a date suggested by Simon Lewis and myself of 1610, and the Great Acceleration, when our environmental impacts grew exponentially in the 1950s and 1960s. In many ways, these arguments detract from the profound change created by recognition of the Anthropocene. This is because it reverses 500 years of scientific discoveries, which have continually moved humans to ever-increasing insignificance. The Copernican revolution of the sixteenth century put the Sun at the centre of the Solar System, downgrading the Earth. Modern cosmology suggests that our Sun is one of 10^{24} stars in the Universe, each one with the potential to have planets. Darwin's nineteenth-century discoveries, and the development of evolutionary studies, established humans as a mere twig on the tree of life, with no special place or origin. In the twenty-first century, adopting the Anthropocene reverses this insignificance: humans are not passive observers of Earth. The role of *Homo sapiens* is central, because the future of the only place where life is known to exist is being determined by our actions. In fact, I would argue that not only has humanity become a geological superpower, it has taken control of evolution itself.

The future of human evolution

Human evolution did not stop with the invention of agriculture; instead, a different set of selection pressures arose. For example, all humans are born with the ability to digest lactose in milk so they can suckle from birth. However, in most humans the enzyme lactase switches off in later life. In the early Holocene Epoch, starting about 9,000 years ago, for small populations in Central Europe, West Africa, and south-west India lactase persisted into adulthood, allowing the digestion of non-human milk and dairy products. Lactase persistence in adults seems to be due to the presence of the key lactase gene. Today, this gene can be found in 80 per cent of Europeans and Americans of European ancestry, while in sub-Saharan Africa and South East Asia its presence is very low. It is absent in the Bantu of South Africa and most Chinese populations. These geographical distributions strongly correlate with the spread of domesticated cattle. Through the early Holocene, dairy farming began and expanded from Central Europe into Western and Northern Europe. Lactose tolerance must have given a significant evolutionary advantage in early agricultural communities. They were probably using milk products as a staple source of energy and as a fall-back food during winter or drought periods. Non-human milk also provides an additional source of Vitamin D, which is essential for normal healthy bones, skin, and blood development. The global spread of both dairy farming and European-derived populations has spread lactase persistence around the globe—though it is interesting to note that the majority of people around the world remain lactose intolerant.

Lactase persistence is just one example of recent changes in human evolution. Another is the appearance of blond(e), fair-skinned peoples (see Box 5). However, the great Stephen Jay Gould declared that he could see no biological change in modern humans over the past 40,000 years. This view is echoed by the geneticist and colleague of mine at University College London, Steve Jones. He argues that modern medicine has halted natural selection, as almost everyone can survive to the age of reproduction, and thus pass on their genetic information. This is, however, a very Western view of our world. First, at this time about 2 billion people do not have access to a modern healthcare system; 1 billion people do not have access to safe, clean drinking water; and 700 million people do not have access to enough food, and go to bed hungry each night. Second, even within a relatively rich developed country there are clear differentiated costs involved in reproduction and health, including access to education, health, contraception, junk food, alcohol, and drugs. Third, as Chris Stringer from the Natural History Museum in London points out, there are large numbers of mutations arising in our genome all the time. One estimate suggests each one of us has fifty mutations that we did not inherit from our parents. With the human population at 7 billion, and set to rise to nearly 10 billion by the middle of the century, that's a lot of mutations—a lot of variation—for evolution to play with. And moreover, increasing globalization

Box 5: Blond Hair and Ocean circulation

The circulation of the North Atlantic Ocean may have helped the very recent evolution of blond fair-skinned people, which has been dated to about 7,000 years ago. The warming effect of the Gulf Stream on Western Europe is so large it means that early agriculturalists could grow crops incredibly far north in countries such as Britain, Norway, and Sweden. In fact remains from seven individuals have been excavated from the Motala archaeological site in southern Sweden and they had the light skin gene variants, SLC24A5 and SLC45A2 and a third gene, HERC2/OCA2, which causes blue eyes and may also contribute to light skin and blond hair. Some of these early settlers were living as far north as the Arctic circle, which on other continents is the middle of the Greenland ice sheet or the northern Alaskan tundra. But there is one major drawback living so far north, and that is the lack of sunlight. Humans need vitamin D, without which children develop many terrible diseases such as rickets, which softens the bones leading to fractures and severe deformity. Vitamin D is made in the skin when it is exposed to ultraviolet light from the sun. This of course was no problem for our ancestors who evolved in Africa, quite the reverse, and dark skin was essential protection from the strong sunlight. However, as our ancestors moved further and further north there was less and less sunlight and less production of Vitamin D. In each generation only those with the lightest skin and hair colour could avoid getting rickets. Because the lighter your skin and hair the more sunlight you can absorb and thus the more Vitamin D you can make. So natural selection favoured two genetic solutions to this problem—evolving pale skin and blond hair that allows UV to be absorbed more efficiently or favouring lactose tolerance to be able to digest the sugars and Vitamin D naturally found in milk. Vitamin D is also found in food including fatty fish species and mushrooms; the former may be why the same selection pressure never applied to the Arctic Inuit. So just think, if it were not for the Gulf Stream and the reliance of the early Scandinavian settlers on their crops, with little or no fish consistently in their diet we would not have real blonds.

means that even previously isolated populations of humans are now interbreeding, producing new genetic combinations.

Many geneticists and anthropologists now argue that culture, far from holding evolution back, is accelerating it. Over the past 10,000 years we have gone through profound changes, from hunter-gatherer to pastoralism to agriculture to urbanization, all of which would have driven evolutionary processes. An example is the human evolutionary reaction to disease. Unsanitary, densely packed urban areas led to the rise of major diseases and parasites—for example, smallpox, cholera, the plague, and yellow fever. Human populations that co-evolved with these diseases built up some genetic resistance. This is shown most dramatically by the devastation of the native population of South America in the sixteenth century, when these diseases were inadvertently introduced. The isolation of this human population from Old World humans for over 10,000 years meant they hadn't evolved immunity to these diseases, leading to over 50 million deaths. Meanwhile about 10 per cent of Europeans had mutations, which provided resistance to smallpox and may also confer resistance to HIV as well, as a fortuitous by-product. In the tropics, the cutting down of forest and widespread use of irrigation led to the spread of malaria across much of the tropical and subtropical area, and the evolution of immunity or resilience in some human populations.

It seems that not only our immune system is constantly evolving, but also our inner ear and eyesight. Recent mutations seem to have changed the genes that produce proteins for the cilia and their membranes in our inner ear, and the bones in our ears, both of which help transmit sonic frequencies. Different mutations with similar results are found in African, European, Chinese, and Japanese populations, suggesting these changes have evolved paralleling the development of different languages and their distinct sounds. Sight may also be undergoing evolutionary pressure, with mutations found in key genes in East Asian populations, changing the photoreceptors in the retina as well as the construction of the inner ear. I also want to speculate as to whether there is a genetic element to the clearly documented reduction in human violence throughout history. Steven Pinker has shown that reactive violence—i.e. individual on individual—can be shown to have reduced in all cultures through time, but now argues that this is due to the refinement and development of better legal and cultural systems in some societies. But what if there is a genetic component, and modern societies naturally select less violent individuals? Chris Stringer speculates that perhaps some of the one hundred or so recent mutations in brain neurotransmitters concerned with mood and demeanour may have been selected to allow us to deal with large population sizes and increased social tensions. However, others have argued that this could be due to environmental causes. In the USA individual-on-individual violence peaked in 1985, ten years after lead was banned from petrol, and has steadily decreased every year since. In the UK individual-on-individual violence peaked in 1990, ten years after lead was banned from petrol, and has steadily decreased every year since. Lead is a well-documented neurotoxin and it could be that it increased violence in the middle of the twentieth century. Others have suggested the rise of violent computer games distracts young males so they do not congregate in large groups outside, and also provides an outlet for young male aggression. Of course both these affects would also lead to changes in brain pattern and structure, which would then be passed on.

Humans are also taking charge of the evolutionary process, both inadvertently and on purpose. Factors such as: the translocation of species, including human populations, to new habitats; the selective removal of particular predators and large mammals; environmental pressure from land-use changes, deforestation, and climate change, are all altering evolutionary outcomes. Add to this the pressures of new pollutants such as antibiotics and other pharmaceuticals, pesticides, heavy metals, microplastics, and novel genetically engineered organisms, and the result is that humans are inadvertently influencing the evolution of many organisms, including ourselves. We are also on the cusp of a human genetic engineering revolution. Already we can counsel parents about harmful DNA mutations that could occur in their children, altering their decisions and changing the future gene pool. Gene therapy will allow DNA to be altered in the germline of an unborn child, changing their evolutionary destiny forever.

If we stray into science fiction we can image a world where humans are fully in control of their own evolution. The late Iain M. Banks, the renowned and brilliant science fiction writer, imagined a human society which was aptly named 'The Culture' that had spread throughout the Universe. Once humans and their AI partners, called 'Minds', had access to the stellar medium, then resources were unlimited and humans controlled their own evolution. Humans within 'The Culture' still have the basic form of *Homo sapiens*, but they also have the ability to change gender whenever they wish, can control their reproductive system, and select particular gene combinations for their children. We are a long way from this, but it does beautifully illustrate the point that we really could be in charge of our evolutionary destiny, unlike the situation today, in which our evolution is only partly intentional and partly continues unplanned, driven by the blind force of natural selection, with unpredictable effects and consequences for the future.

Homo dominatus

Human evolution is the story of how we came to walk upright and then became progressively smarter. All these major evolutionary changes occurred in Africa. It seems that without the unique impact of tectonics, celestial mechanics, and global climate variations on the landscape of East Africa, human evolution would never have occurred. With the extinction of the Neanderthals about 40,000 years ago, *H. sapiens* became the last surviving hominin. With no natural weapons or defences, we dispersed out of Africa as the apex predator and populated all the continents except Antarctica. The complex, flexible social human brain facilitated the development of cumulative culture. This led to the first appearance of agriculture at the end of the last ice age. Throughout the next 10,000 years, agriculture spread to almost every region of the world, steadily replacing hunter-gathering as the major economic mainstay, allowing urban centres and city-states to develop. The intense competition between small nation states in Europe, from the eleventh century onwards, created the perfect centre for technological innovation, particularly of weapons. By the fifteenth century, European powers had outstripped the Chinese and Ottoman Empires and begun the great age of exploration. Annexation and depopulation of the Americas in the sixteenth century provided wealth and produce, allowing the European population to grow beyond its own agricultural productivity limits. Continued competition between the European powers, and capital from their empires, led to the industrial revolution of the eighteenth century. Post-World War II prosperity created the technological revolution and the Great Acceleration of environmental impacts. We have now entered what many see as the fourth revolution—the information revolution. All these revolutions have been driven by our cumulative culture and our ultrasociality. It has also allowed the human population to expand to over 7.5 billion people and it is set to rise to nearly 10 billion people by the middle of the century. We have influenced almost every part of the Earth system. We are changing the global environmental and evolutionary trajectory of the Earth. We are also both inadvertently and on purpose changing our own evolution. Humanity has become in effect a geological superpower.

What I hope this book has demonstrated is that we were in many ways created by rapid environmental changes that occurred in East Africa, evolving a large social brain with intellectual flexibility and group dynamics to deal with rapid and surprising changes. The questions are: can we as a species cope with the global environmental changes we ourselves have inadvertently caused? Can humans, the only ultrasocial creature, knowingly self-organize nearly 10 billion individuals in the future to mitigate and/or adapt to the destruction of our environment? Or are we limited by our evolutionary history? Because we are the only known species that can predict the results of our own actions and truly understand how we could positively manage the Earth's environment for the benefit of ourselves and all other species. We have named ourselves *Homo sapiens*—'wise person'. It is clear that as a species we have taken control over the environmental and biological future of

Earth. Maybe *Homo dominatus*—which translates as 'dominant person' or 'tyrannical person'—would be more apt.

For me, the flexibility of the human brain and the fact that many of its structures are determined not by genetics, but by the environment and society into which we are born, means that we can learn new ways of living very quickly. Each generation of humanity can therefore adapt to the new environmental, social, and political challenges facing us. Maybe in this century, if we take our impacts on the Earth system seriously, we will earn our species name 'wise'. Whatever the future holds, we have come a long way from our cradle in East Africa.

REFERENCES AND FURTHER READING

CHAPTER 1: INTRODUCTION

Books

Charlesworth, B. and Charlesworth, D. 2003. *Evolution: A Very Short Introduction*, Oxford University Press, Oxford, p. 145.

Colling, A. (ed.) 1997. *The Earth and Life: The Dynamic Earth*, Open University Worldwide, Milton Keynes, p. 256.

Langmuir, C. H. and Broecker, W. 2012. *How to Build a Habitable Planet: The Story of Earth from the Big Bang to Humankind*, revised and expanded edition, Princeton University Press, Princeton, NJ, p. 720.

Lenton, T. and Watson, A. 2011. *Revolutions that Made the Earth*, Oxford University Press, Oxford, p. 423.

Lockwood, C. 2008. *The Human Story: Where We Come From and How We Evolved*, Natural History Museum, London, p. 111.

Zalasiewicz, J. and Williams, M. 2012. *The Goldilocks Planet: The 4 billion Year Story of Earth's Climate*, Oxford University Press, Oxford, p. 336.

CHAPTER 2: EARLY HUMAN EVOLUTION

Books

Grine, F., Leakey, R., and Fleagle, J. (eds) 2009. *The First Humans: Origins and Early Evolution of the Genus Homo*, Springer, New York, p. 218.

Leakey, R. and Lewin, R. 1992. *Origins Reconsidered: In Search of What Makes Us Human*, Little, Brown and Company, London, p. 375.

Lewin, R. and Foley, R.A. 2004. *Principles of Human Evolution*, John Wiley & Sons, London.

Lockwood, C. 2008. *The Human Story: Where We Come From and How We Evolved*, Natural History Museum, London, p. 111.

Reader, J. 2011. *Missing Links: In Search of Human Origins*, 3rd edition, Oxford University Press, New York.

Roberts, A. 2014. *The Incredible Unlikeness of Being: Evolution and the Making of Us*, Heron Books, London, p. 392.

Sponheimer, M., Lee-Thorp, J.A., Reed, K.E., and Ungar, P. (eds) 2013. *Early Hominin paleoecology*, University of Colorado Press, Boulder, CO, p. 470.

Stringer, C. 2012. *The Origin of Our Species*, Penguin, London, p. 333.

Wood, B. 2005. *Human Evolution: A Very Short Introduction*, Oxford University Press, Oxford, p. 131.

Journals and chapters

Aiello, L.C. and Andrews, P. 2000. The Australopithecines in review. *Human Evolution* 15: 17–38.

Aiello, L.C. and Key, C. 2012. Energetic consequences of being an *erectus* female. *American Journal of Human Biology* 14(5): 551–65.

Antón, S.C. 2003. Natural history of *Homo erectus*. *American Journal of Physical Anthropology* 122: 126–70.

Antón, S.C., Aiello, L.C., and Potts, R. 2014. Evolution of early *Homo*: An integrated biological perspective. *Science* 345(6192).

Bentley-Condit, V. and Smith, E.O. 2010. Animal tool use: current definitions and an updated comprehensive catalog. *Behaviour* 147: 185.

Bramble, D.M. and Lieberman, D.E. 2004. Endurance running and the Evolution of Homo. *Nature* 432: 345–52.

Brumm, A., van den Bergh, G.D., Storey, M. et al. 2016. Age and context of the oldest known hominin fossils from Flores. *Nature* 534: 249–53.

Carmody, R.N. and Wrangham, R. 2009. The energetic significance of cooking. *Journal of Human Evolution* 57(4): 379–91.

Collard, M. and Wood, B.A. 2007. Defining the genus Homo, in Henke, W., Rothe, H., Tattersall, I., (eds) *Handbook of Paleoanthropology*, Springer, Heidelberg, pp. 1575–610.

Guy, F., Lieberman, D.E., Pilbeam, D. et al. 2005. Morphological affinities of the *Sahelanthropus tchadensis* (Late Miocene hominid from Chad) cranium. *Proceedings of the National Academy of Sciences* 102: 18836–41.

Haile-Selassie, Y., Suwa, G., and White, T.D. 2004. Late Miocene teeth from Middle Awash, Ethiopia, and early hominid dental evolution. *Science* 303: 1503–5.

Haile-Selassie, Y., Saylor, B.Z., Deino, A. et al. 2012. A new hominin foot from Ethiopia shows multiple Pliocene bipedal adaptations. *Nature* 483: 565–9.

Haile-Selassie, Y., Gibert, L., Melillo, S.M. et al. 2015. New species from Ethiopia further expands Middle Pliocene hominin diversity. *Nature* 521: 483–8.

Harmand S., Lewis, J.E., Feibel, C. et al. 2015. 3.3-million-year-old stone tools from Lomekwi 3, West Turkana, Kenya. *Nature* 521: 310–15.

Hublin J-J., et al. 2017. New fossils from Jebel Irhoud, Morocco and the pan-African origin of *Homo sapiens*. *Nature* 546: 289–92.

Lordkipanidze, D., Ponce de León, M.S., Margvelashvili, A. et al. 2013. A complete skull from Dmanisi, Georgia, and the evolutionary biology of early *Homo*. *Science* 342: 326–31.

Lovejoy, C. Owen. 2005. The natural history of human gait and posture. Part 1. Spine and pelvis. *Gait & Posture*. 21(1): 113–24.

Maslin, M.A., Brierley, C.M., Milner, A.M. et al. 2014. East African climate pulses and early human evolution. *Quaternary Science Reviews* 101: 1–17.

McPherron, S.P., Alemseged, Z., Marean, C.W. et al. 2010. Evidence for stone-tool-assisted consumption of animal tissues before 3.39 million years ago at Dikika, Ethiopia. *Nature* 466: 857–60.

Roach, N.T., Venkadesan, N., Rainbow, M., and Lieberman, D.E. 2013. Elastic energy storage in the shoulder and the evolution of high-speed throwing in Homo. *Nature* 498: 483–7.

Senut, B., Pickford, M., Gommery, D. et al. 2001. First hominid from the Miocene (Lukeino Formation, Kenya). *Earth and Planetary Science Letters* 332: 137–44.

Spoor, F., Gunz, P., Neubauer, S. et al. 2015. Reconstructed *Homo habilis* type OH 7 suggests deep-rooted species diversity in early *Homo*. *Nature* 519: 83–6.

Sutikna, T., Tocheri, M.W., Morwood, M.J. et al. 2016. Revised stratigraphy and chronology for *Homo floresiensis* at Liang Bua in Indonesia. *Nature* 532: 366–9.

Villmoare, B., Kimbel, W.H., Seyoum, C. et al. 2015. Early *Homo* at 2.8 Ma from Ledi-Geraru, Afar, Ethiopia. *Science* 347: 1352–5.

White, T.D., Asfaw, B., Beyene, Y. et al. 2009. *Ardipithecus ramidus* and the paleobiology of early hominids. *Science* 64: 75–86.

Wood, B. 2002. Palaeoanthropology: Hominid revelations from Chad. *Nature* 418: 133–5.

Wood, B. 2014. Fifty years after *Homo habilis*. *Nature* 508: 31–3.

Wood, B. and Collard, M. 1999. The changing face of the genus *Homo*. *Evolutionary Anthropology* 8(6): 197–207.

Wood, B. and Strait, D. 2004. Patterns of resource use in early *Homo* and *Paranthropus*. *Journal of Human Evolution* 46: 119–62.

Zhu, Z. et al. 2018. Hominin occupation of the Chinese Loess Plateau since about 2.1 million years ago, *Nature* https://doi.org/10.1038/s41586-018-0299-4.

Zink, K.D. and Lieberman, D.E. 2016. Impact of meat and Lower Palaeolithic food processing techniques on chewing in humans. *Nature* 531: 500–3.

CHAPTER 3: TECTONICS AND CLIMATE

Books

Maslin, M. 2013. *Climate: A Very Short Introduction*, Oxford University Press, Oxford, p. 159.

Ruddiman, W.F. 2007. *Earth's Climate: Past and Future*, 2nd edition, W.H. Freeman, New York, p. 480.

Van Andel, T. 1994. *New Views on an Old Planet: A History of Geological Change*, Cambridge University Press, Cambridge, p. 458.

Journals

Cane, M.A. and Molnar, P. 2001. Closing of the Indonesian seaway as a precursor to east African aridification around 3–4 million years ago. *Nature* 411: 157–62.

DeConto, R.M., Pollard, D., Wilson, P.A. et al. 2008. Thresholds for Cenozoic bipolar glaciation. *Nature* 455: 652–6.

Haug, G.H. and Tiedemann, R. 1998. Effect of the formation of the Isthmus of Panama on Atlantic Ocean thermohaline circulation. *Nature* 393: 673–5.

Maslin, M.A. and Christensen, B. 2007. Tectonics, orbital forcing, global climate change, and human evolution in Africa. *Journal of Human Evolution* 53(5): 443–64.

Raymo, M.E. 1994. The initiation of Northern Hemisphere glaciation. *Annual Review of Earth and Planetary Sciences* 22: 353–83.

Ruddiman, W.F. and Raymo, M.E. 1988. Northern Hemisphere climate regimes during the past 3 Ma: Possible tectonic connections. *Philosophical Transactions of the Royal Society B* 318: 411–30.

Sepulchre, P., Ramstein, G., Fluteau, F., and Schuster, M. 2006. Tectonic uplift and Eastern Africa aridification. *Science* 313: 1419–23.

Trauth, M.H., Maslin, M.A., Deino, A., and Strecker, M.R. 2005. Late Cenozoic moisture history of East Africa. *Science* 309: 2051–3.

Trauth, M.H., Maslin, M.A., Deino, A.L. et al. 2007. High- and low-latitude controls and East African forcing of Plio-Pleistocene East African climate and early human evolution. *Journal of Human Evolution* 53: 475–86.

Zachos, J.C., Pagani, M., Sloan, L. et al. 2001. Trends, rhythms and aberrations in global climate 65 Ma to present. *Science* 292: 686–93.

CHAPTER 4: CRADLE OF HUMANITY

Books

Lewin, R. and Foley, R.A. 2004. *Principles of Human Evolution*, John Wiley & Sons, London.

Journals

Arsuaga, J.L. 2010. Terrestrial apes and phylogenetic trees. *Proceedings of the National Academy of Sciences* 107: 8910–17.

Barboni, D. 2014. Vegetation of Northern Tanzania during the Plio-Pleistocene: A synthesis of the paleobotanical evidences from Laetoli, Olduvai, and Peninj hominin sites. *Quaternary International* 322: 264–76.

Blisniuk, P. and Strecker, M.R. 1990. Asymmetric rift-basin development in the central Kenya Rift. *Terra Abstracts* 2: 51.

Bonnefille, R. 2010. Cenozoic vegetation, climate changes and hominid evolution in tropical Africa. *Global and Planetary Change* 72: 390–411.

Brachert, T.C., Brugmann, G.B., Mertz, D.F. et al. 2010. Stable isotope variation in tooth enamel from Neogene hippopotamids: Monitor of meso and global climate and rift dynamics on the Albertine Rift, Uganda. *International Journal of Earth Sciences* 99(7): 1663–75.

Cerling, T.E. 2014. Stable isotope evidence for hominin environments in Africa. *Treatise on Geochemistry*, 2nd edition <http://dx.doi.org/10.1016/B978-0-08-095975-7.01213-4>.

Couvreur, T.L.P., Chatrou, L.W., Sosef, M.S.M., and Richardson, J.E. 2008. Molecular phylogenetics reveal multiple tertiary vicariance origins of African rain forest trees. *BMC Biology* 6: 54.

Crompton, R., Vereecke, E., and Thorpe, S. 2008. Locomotion and posture from the common hominoid ancestor to fully modern hominins, with special reference to the last common panin/hominin ancestor. *Journal of Anatomy* 212: 501–43.

Crompton, R., Sellers, W., and Thorpe, S. 2010. Arboreality, terrestriality and bipedalism. *Philosophical Transactions of the Royal Society B* 365: 3301–14.

Ebinger, C.J., Yemane, T., Harding, D.J. et al. 2000. Rift deflection, migration, and propagation: Linkage of the Ethiopian and Eastern rifts. *Geological Society of America Bulletin* 112: 163–76.

Feakins, S.J., deMenocal, P.B., and Eglinton, T.I. 2005. Biomarker records of late Neogene changes in northeast African vegetation. *Geology* 33: 977–80.

Feakins, S.J., Levin, N.E., Liddy, H.M. et al. 2013. Northeast African vegetation change over 12 m.y. *Geology* 41: 295–8.

Feibel, C.S. 2011. Geological history of the Turkana Basin. *Evolutionary Anthropology: Issues, News, and Reviews* 20: 206–16.

Forster, A. and Gleadow, J.W. 1996. Structural framework and denudation history of the flanks of the Kenya and Anza Rifts, East Africa. *Tectonics* 15: 258–71.

Foster, A., Ebinger, C., Mbede, E., and Rex, D. 1997. Tectonic development of the northern Tanzanian sector of the east African rift system. *Journal of the Geological Society* 154: 689–700.

Levin, N.E., Quade, J., Simpson, S.W. et al. 2004. Isotopic evidence for Plio-Pleistocene environmental change at Gona, Ethiopia. *Earth and Planetary Science Letters* 219: 93–110.

Levin N.E. 2015. Environment and climate of early human evolution. *Annual Review of Earth and Planetary Sciences* 43: 405–29.

Maslin, M.A. and Christensen, B. 2007. Tectonics, orbital forcing, global climate change, and human evolution in Africa. *Journal of Human Evolution* 53(5): 443–64.

Maslin, M.A. and Thomas , E. 2003. Balancing the deglacial global carbon budget: The hydrate factor. *Quaternary Science Reviews* 22(15–17): 1729–36).

Maslin, M.A., Pancost, R., Wilson, K.E. et al. 2012. Three and half million year history of moisture availability of South West Africa: Evidence from ODP site 1085 biomarker records. *Palaeogeography, Palaeoclimatology, Palaeoecology* 317–18: 41–7.

Pik, R., Marty, B., Carignan, J. et al. 2008. Timing of East African Rift development in Southern Ethiopia: Implication for mantle plume activity and evolution of topography. *Geology* 36: 167–70.

Prömmel, K., Cubasch, U., and Kasper, F. 2013. A regional climate model study of the impact of tectonic and orbital forcing on African precipitation and vegetation. *Palaeogeography, Palaeoclimatology, Palaeoecology* 369: 154–62.

Roberts, A. and Thorpe, S. 2014. Challenges to human uniqueness: Bipedalism, birth and brains. *Journal of Zoology* 292: 281–9.

Sepulchre, P., Ramstein, G., Fluteau, F., and Schuster, M. 2006. Tectonic uplift and Eastern Africa aridification. *Science* 313: 1419–23.

Strecker, M.R., Blisniuk, P.M., and Eisbacher, G.H. 1990. Rotation of extension direction in the central Kenya Rift. *Geology* 18: 299–302.

Underwood, C.J., King, C., and Steurbaut, E. 2013. Eocene initiation of Nile drainage due to East African uplift. *Palaeogeography, Palaeoclimatology, Palaeoecology* 392: 138–45.

Wichura, H., Bousquet, R., Oberhänsli, R. et al. 2010. Evidence for Mid-Miocene uplift of the East African Plateau. *Geology* 38: 543–6.

Wynn, J.G. 2004. Influence of Plio-Pleistocene aridification on human evolution: Evidence from Paleosols of the Turkana Basin, Kenya. *American Journal of Physical Anthropology* 123: 106–18.

CHAPTER 5: GLOBAL CLIMATE CHANGE

Books

Alley, R.B. 2002. *The Two-Mile Time Machine: Ice Cores, Abrupt Climate Change, and Our Future*, Princeton University Press, Princeton, NJ, p. 240.

Fagan, B. (ed.) 2009. *The Complete Ice Age: How Climate Change Shaped the World*, Thames and Hudson, London, p. 240.

Lowe, J.J. and Walker, M. 1997. *Reconstructing Quaternary Environments*, 2nd edition, Prentice Hall, London, p. 472.

Maslin, M. 2013. *Climate: A Very Short Introduction*, Oxford University Press, Oxford, p. 159.

Ruddiman, W.F. 2007. *Earth's Climate: Past and Future*, 2nd edition, W.H. Freeman, New York, p. 480.

Wilson, R.C.L., Drury, S.A., and Chapman, J.L. 2003. *The Great Ice Age: Climate Change and Life*, Routledge, London.

Journals and chapters

Berger, W.H. and Jansen, E. 1994. Mid-Pleistocene climate shift: The Nansen connection, in: Johannessen, O.M., Muench, R.D., and Overland, J.E. (eds), *The Polar Oceans and Their Role in Shaping the Global Environment*, American Geophysical Union, Washington, DC, pp. 295–311.

Bickert, T.A., Haug, G.H., and Tiedemann, R. 2004. Late Neogene benthic stable isotope record of Ocean Drilling Program Site 999: Implications for Caribbean paleoceanography, organic carbon burial, and the Messinian Salinity Crisis. *Paleoceanography* 19(1023).

Brierley, C., Fedorov, A.V., Lui, Z. et al. 2010. Greatly expanded tropical warm pool and weakened Hadley circulation in the Early Pliocene. *Science* 323: 1714–18.

Brown, N.J., Newell, C.A., Stanley, S. et al. 2011. Independent and parallel recruitment of preexisting mechanisms underlying C4 photosynthesis. *Science* 332: 1436–9.

Clemens, S.C., Murray, D.W., and Prell, W.L. 1996. Nonstationary phase of the Plio-Pleistocene Asian monsoon. *Science* 274: 943–8.

Edwards, E.J., Osborne, C.P., Strömberg, C.A.E., Smith, S.A., and C4 Grasses Consortium. 2010. The origins of C4 grasslands: Integrating evolutionary and ecosystem science. *Science* 331: 587–91.

Fedorov, A.V., Brierley, C., Lawrence, K.T. et al. 2013. Patterns and mechanisms of early Pliocene warmth. *Nature* 496: 43–9.

Imbrie, J., Boyle, E., Clemens, S. et al. 1992. On the structure and origin of major glaciation cycles. 1. Linear responses to Milankovitch forcing. *Paleoceanography* 7: 701–38.

Keigwin, L.D. 1982. Pliocene paleoceanography of the Caribbean and east Pacific: Role of Panama uplift in late Neogene times. *Science* 217: 350–3.

Keller, G., Zenker, C.E., and Stone, S.M. 1989. Late Neogene history of the Pacific-Caribbean gateway. *Journal of South American Earth Sciences* 2: 73–108.

Maslin, M.A., Li, X.S., Loutre, M.F., and Berger, A. 1998. The contribution of orbital forcing to the progressive intensification of Northern Hemisphere Glaciation. *Quaternary Science Reviews* 17: 411–26.

Maslin, M.A., Seidov, D., and Lowe, J. 2001. Synthesis of the nature and causes of sudden climate transitions during the Quaternary, in Seidov, D., Haupt, B.J., and Maslin, M.A. (eds), *The Oceans and Rapid Climate Change: Past, Present and Future*, Geophysical Monograph Series, American Geophysical Union, Washington, DC, pp. 9–52.

McClymont, E.L. and Rosell-Melé, A. 2005. Links between the onset of modern walker circulation and the mid-Pleistocene climate transition. *Geology* 33: 389–92.

Mudelsee, M. and Stattegger, K. 1997. Exploring the structure of the mid-Pleistocene revolution with advance methods of time-series analysis. *Geologische Rundschau* 86: 499–511.

Murphy, L.N., Kirk-Davidoff, D.B., Mahowald, N., and Otto-Bliesner, B.L. 2009. A numerical study of the climate response to lowered Mediterranean Sea level during the Messinian Salinity Crisis. *Palaeogeography, Palaeoclimatology, Palaeoecology* 279: 41–59.

Ravelo, C., Andreasen, D., Lyle, M. et al. 2004. Regional climate shifts caused by gradual global cooling in the Pliocene epoch. *Nature* 429: 263–7.

Raymo, M.E. 1991. Geochemical evidence supporting T.C. Chamberlin's theory of glaciation. *Geology* 19: 344–7.

Raymo, M.E., 1994. The initiation of Northern Hemisphere glaciation. *Annual Review of Earth and Planetary Sciences* 22: 353–83.

Roveri, M., Lugli, S., Manzi, V., and Schreiber, B.C. 2008. The Messinian Sicilian stratigraphy revisited: New insights for the Messinian Salinity Crisis. *Terra Nova* 20: 483–8.

Schneck, R., Micheels, A., and Mosbrugger, V. 2010. Climate modelling sensitivity experiments for the Messinian Salinity Crisis. *Palaeogeography, Palaeoclimatology, Palaeoecology* 286: 149–63.

Ségalen, L., Lee-Thorp J.A., and Cerling T. 2007. Timing of C4 grass expansion across sub-Saharan Africa. *Journal of Human Evolution* 53: 549–59.

Tipple B.J. and Pagani M. 2007. The early origins of C_4 photosynthesis. *Annual Review of Earth and Planetary Sciences* 35: 435–61.

Trauth, M.H., Larrasoaña, J.C., and Mudelsee, M. 2009. Trends, rhythms and events in Plio-Pleistocene African climate. *Quaternary Science Reviews* 28: 399–411.

Zachos, J.C., Pagani, M., Sloan, L. et al. 2001. Trends, rhythms and aberrations in global climate 65 Ma to present. *Science* 292: 686–93.

CHAPTER 6: CELESTIAL MECHANICS

Books

Maslin, M. 2013. *Climate: A Very Short Introduction*, Oxford University Press, Oxford, p. 159.

Ruddiman, W.F. 2007. *Earth's Climate: Past and Future*, 2nd edition, W.H. Freeman, New York, p. 480.

Wilson, R.C.L., Drury, S.A., and Chapman, J.L. 2003. *The Great Ice Age: Climate Change and Life*, Routledge, London.

Journals and chapters

Berger, A., Loutre, M.F., and Mélice, J.L. 2006. Equatorial insolation: From precession harmonics to eccentricity frequencies. *Climate of the Past* 2: 131–6.

Clement, A.C., Hall, A., and Broccoli, A.J. 2004. The importance of precessional signals in the tropical climate. *Climate Dynamics* 22: 327–41.

Deino, A.L., Kingston, J.D., Glen, J.M. et al. 2006. Precessional forcing of lacustrine sedimentation in the late Cenozoic Chemeron Basin, Central Kenya Rift, and calibration of the Gauss/Matuyama boundary. *Earth and Planetary Science Letters* 247: 41–60.

deMenocal, P. 1995. Plio-Pleistocene African climate. *Science* 270: 53–9.

deMenocal, P. 2004. African climate change and faunal evolution during the Pliocene-Pleistocene. *Earth and Planetary Science Letters* 220: 3–24.

Denison, S., Maslin, M.A., Boot, C. et al. 2005. Precession-forced changes in South West African vegetation during marine oxygen isotope stages 100 and 101. *Palaeogeography, Palaeoclimatology, Palaeoecology* 220: 375–86.

Donges, J.F., Donner, R.V., Trauth, M.H. et al. 2011. Nonlinear detection of paleoclimate-variability transitions possibly related to human evolution. *Proceedings of the National Academy of Sciences* 108: 20422–7.

Hays, J.D., Imbrie, J., and Shackleton, N.J. 1976. Variations in the Earth's orbit: Pacemaker of the Ice Ages. *Science* 194: 1121–32.

Hopley, P.J. and Maslin, M.A. 2010. Climate-averaging of terrestrial faunas: An example from the Plio-Pleistocene of South Africa. *Palaeobiology* 36: 32–50.

Hopley, P.J., Marshall, J.D., Weedon, G.P. et al. 2007. Orbital forcing and the spread of C4 grasses in the late Neogene: Stable isotope evidence from South African speleothems. *Journal of Human Evolution* 53: 620–34.

Joordens, J.C.A., Vonhof, H.B., Feibel, C.S. et al. 2011. An astronomically-tuned climate framework for hominins in the Turkana Basin. *Earth and Planetary Science Letters* 307: 1–8.

Junginger, A. and Trauth, M.H. 2013. Hydrological constraints of paleo-Lake Suguta in the Northern Kenya Rift during the African Humid Period (15–5 ka BP). *Global and Planetary Change* 111: 174–88.

Junginger, A., Roller, S., Olaka, L., and Trauth, M.H. 2014. The effect of solar irradiation changes on water levels in the paleo-Lake Suguta, Northern Kenya Rift, during the late Pleistocene African Humid Period (15–5 ka BP). *Palaeogeography, Palaeoclimatology, Palaeoecology* 396: 1–16.

Kappelman, J. 1993. The attraction of paleomagnetism. *Evolutionary Anthropology* 2(3): 89–99.

Kingston, J.D., Deino, A.L., Edgar, R.K., and Hill, A. 2007. Astronomically forced climate change in the Kenyan Rift Valley 2.7–2.55 Ma: Implications for the evolution of early hominin ecosystems. *Journal of Human Evolution* 53: 487–503.

Larrasoaña, J.C., Roberts, A.P., Rohling, E.J. et al. 2003. Three million years of monsoon variability over the northern Sahara. *Climate Dynamics* 21: 689–98.

Larrasoaña, J.C., Roberts, A.P., and Rohling E.J. 2013. Dynamics of green Sahara periods and their role in hominin evolution. *PLoS ONE* 8(10): e76514.

Magill, C.R., Ashley, G.M., and Freeman, K. 2013. Ecosystem variability and early human habitats in eastern Africa. *Proceedings of the National Academy of Sciences* 110: 1167–74.

Maslin, M.A., Li, X.S., Loutre, M.F., and Berger, A. 1998. The contribution of orbital forcing to the progressive intensification of Northern Hemisphere Glaciation. *Quaternary Science Reviews* 17: 411–26.

Maslin, M.A., Seidov, D., and Lowe, J. 2001. Synthesis of the nature and causes of sudden climate transitions during the Quaternary, in Seidov, D., Haupt, B.J., and Maslin, M.A. (eds), *The Oceans and Rapid Climate Change: Past, Present and Future*, Geophysical Monograph Series 126, American Geophysical Union, Washington, DC, pp. 9–52.

Maslin, M.A., Brierley, C.M., Milner, A.M. et al. 2014. East African climate pulses and early human evolution. *Quaternary Science Reviews* 101: 1–17.

Milankovitch, M.M. 1949. Kanon der Erdbestrahlung und seine Anwendung auf das Eiszeitenproblem. Royal Serbian Sciences, Special Publication 132, Section of Mathematical and Natural Sciences, 33, Belgrade, p. 633 (Canon of Insolation and the Ice Age Problem, English translation by Israel Program for Scientific Translation and published for the US Department of Commerce and the National Science Foundation, Washington, DC, 1969).

Olaka, L.A., Odada, E.O., Trauth, M.H., and Olago, D.O. 2010. The sensitivity of East African rift lakes to climate fluctuations. *Journal of Paleolimnology* 44: 629–44.

Trauth, M.H., Maslin, M.A., Deino, A., and Strecker, M.R. 2005. Late Cenozoic moisture history of East Africa. *Science* 309: 2051–3.

Trauth, M.H., Maslin, M.A., Deino, A.L. et al. 2007. High- and low-latitude controls and East African forcing of Plio-Pleistocene East African climate and early human evolution. *Journal of Human Evolution* 53: 475–86.

Trauth, M.H., Maslin, M.A., Bergner, A.G.N. et al. 2010. Human evolution and migration in a variable environment: The amplifier lakes of East Africa. *Quaternary Science Reviews* 29: 2981–8.

Trauth, M.H., Bergner, A.G.N., Foerster, V. et al. 2015. Episodes of environmental stability vs instability in Late Cenozoic lake records of Eastern Africa. *Journal of Human Evolution*. 87: 21–31

Veldhuis, D., Kjærgaard, P.C., and Maslin, M. 2014. Human evolution: Theories and progress, in Smith, C. (ed.) *Encyclopedia of Global Archaeology*, Springer, New York, pp. 3520–32.

CHAPTER 7: AFRICAN CLIMATE PULSES

Books

Darwin, C. 1859. *On the Origin of Species by Means of Natural Selection, or the Preservation of Favoured Races in the Struggle for Life*, John Murray, London.

Darwin, C. 1871. *The Descent of Man and Selection in Relation to Sex, Volume 1*, John Murray, London.

Lamarck, J.B. 1809. *Philosophie Zoologique (Zoological Philosophy: Exposition with Regard to the Natural History of Animals) Volume 1*, Museum d'Histoire Naturelle (Jardin des Plantes), Paris.

Journals and chapters

Barnosky, A.D. 2001. Distinguishing the effects of the Red Queen and court jester on Miocene mammal evolution in the northern Rocky Mountains. *Journal of Vertebrate Paleontology* 21: 172–85.

Bergner, A.G.N., Strecker, M.R., Trauth, M.H. et al. 2009. Tectonic versus climate influences on the evolution of the lakes in the Central Kenya Rift. *Quaternary Science Reviews* 28: 2804–16.

deMenocal, P. 1995. Plio-Pleistocene African climate. *Science* 270: 53–9.

Faith, J.T. and Behrensmeyer, A.K. 2013. Climate change and faunal turnover: Testing the mechanics of the turnover-pulse hypothesis with South African fossil data. *Paleobiology* 39: 609–27.

Flinn, M.V., Geary, D.C., and Ward, C.V. 2005. Ecological dominance, social competition, and coalitionary arms races: Why humans evolved extraordinary intelligence. *Evolution and Human Behavior* 26: 10–46.

Foley, R.A. 1994. Speciation, extinction and climatic change in hominid evolution. *Journal of Human Evolution* 26(4): 275–89.

Foley, R.A. 2002. Adaptive radiations and dispersals in hominin evolutionary ecology. *Evolutionary Anthropology* 51(s1): 32–7.

Grove, M. 2011. Change and variability in Plio-Pleistocene climates: Modelling the hominin response. *Journal of Archaeological Science* 38: 3038–47.

Grove, M. 1 2011. Speciation, diversity, and Mode 1 technologies: The impact of variability selection. *Journal of Human Evolution* 61: 306–19.

Hopf, F.A., Valone, T.J., and Brown, J.H. 1993. Competition theory and the structure of ecological communities. *Evolutionary Ecology* 7: 142–54.

Kingston, J.D. 2007. Shifting adaptive landscapes: Progress and challenges in reconstructing early hominid environments. *American Journal of Physical Anthropology: Yearbook of Physical Anthropology* 134, Issue Supplement 45: 20–58.

Kingston, J.D., Marino, B.D., and Hill, A. 1994. Isotopic evidence for Neogene hominid paleoenvironments in the Kenya Rift Valley. *Science* 264: 955–9.

Kjærgaard, P.C. 2011. 'Hurrah for the missing link!': A history of apes, ancestors and a crucial piece of evidence. *Notes and Records of the Royal Society* 65(1): 83–98.

Maslin, M.A. and Trauth, M.H. 2009. Plio-Pleistocene East African pulsed climate variability and its influence on early human evolution, in Grine, F.E., Leakey, R.E., and Fleagle, J.G. (eds),

The First Humans: Origins of the Genus Homo, Springer, New York, pp. 151–8.

Maslin, M.A., Brierley, C., Milner, A. et al. 2014. East African climate pulses and early human evolution. *Quaternary Science Reviews* 101: 1–17.

Maslin, M.A., Shultz, S., and Trauth, M. 2015. A synthesis of the theories and concepts of early human evolution. *Philosophical Transactions of the Royal Society B* 370: 20140064.

Potts, R. 1996. Evolution and climatic variability. *Science* 273: 922–3.

Potts, R. 1998. Environmental hypothesis of hominin evolution. *Yearbook of Physical Anthropology* 41: 93–136.

Potts, R. 1999. Variability selection in hominid evolution. *Evolutionary Anthropology* 1999: 81–96.

Potts, R. 2013. Hominin evolution in settings of strong environmental variability. *Quaternary Science Reviews* 73: 1–13.

Reed, K.E. 1997. Early hominid evolution and ecological change through the African Plio-Pleistocene. *Journal of Human Evolution* 32: 289–322.

Reed, K.E. and Fish, J.L. 2005. Tropical and temperate seasonal influences on human evolution, in Brockman, D. and van Schaik, C. (eds), *Seasonality in Primates*, Cambridge University Press, Cambridge, pp. 491–520.

Reed, K.E. and Russak, S.M. 2009. Tracking ecological change in relation to the emergence of Homo near the Plio-Pleistocene boundary, in Grine, F.E., Leakey, R.E., and Fleagle, J.G. (eds), *The First Humans: Origins of the Genus Homo*, Springer, New York, pp. 159–71.

Shultz, S. and Maslin, M.A. 2013. Early human speciation, brain expansion and dispersal influenced by African climate pulses. *PLoS ONE* 8(10): e76750.

Shultz, S. and Maslin, M. 2013. Early human speciation, dispersal and brain expansion forced by East African climate pulses. *PLoS ONE* 8(10): e76750.

Shultz, S., Nelson, E., and Dunbar, R.I.M. 2012. Hominin cognitive evolution: Identifying patterns and processes in the fossil and archaeological record. *Philosophical Transactions of the Royal Society B* 367: 2130–40.

Vrba, E.S. 1985. Environment and evolution: Alternative causes of the temporal distribution of evolutionary events. *South African Journal of Science* 81: 229–36.

Vrba, E.S. 1988. Late Pliocene climatic events and hominid evolution, in Grine, F. (ed.), *Evolutionary History of the 'Robust' Australopithecines*, De Gruyter, Berlin, pp. 405–26.

Vrba, E.S. 1995. The fossil record of African antelopes (Mammalia, Bovidae) in relation to human evolution and paleoclimate, in Vrba, E.S., Denton, G., Burckle, L., and Partridge, T. (eds), *Paleoclimate and Evolution with Emphasis on Human Origins*, Yale University Press, New Haven, CT, pp. 385–424.

Vrba, E.S. 2000. Major features of Neogene mammalian evolution in Africa, in Partridge, T.C. and Maud, R.R. (eds), *The Cenozoic of Southern Africa*, Oxford University Press, New York, pp. 277–304.

Wilson, K.E., Maslin, M.A., Leng, M.J. et al. 2014. East African lake evidence for Pliocene millennial-scale climate variability. *Geology* 42(11): 955–8.

CHAPTER 8: THE SOCIAL BRAIN

Books

Diamond, J. 1992. *The Rise and Fall of the Third Chimpanzee*, Vintage, London, pp. 60–1.

Dixson, A.F. 2012. *Primate Sexuality*, 2nd edition, Oxford University Press, Oxford, p. 785.

Dunbar, R. 2014. *Human Evolution: A Pelican Introduction*, Penguin, London, p. 415.

Gamble, C., Gowlett, J., and Dunbar, R. 2014. *Think Big: How the Evolution of Social Life Shaped the Human Mind*, Thames and Hudson, London, p. 224.

Hood, B. 2014. *The Domesticated Brain: A Pelican Introduction*, Penguin, London, p. 336.

Journals and chapters

Agusti, J. and Lordkipanidze, D. 2011. How 'African' was the early human dispersal out of Africa? *Quaternary Science Reviews* 30: 1338–42.

Antón, S.C. and Swisher, C.C. 2004. Early dispersals of *Homo* from Africa. *Annual Review of Anthropology* 33: 271–96.

Armitage, S.J., Jasim, S.A., Marks, A.E. et al. 2011. The southern route 'out of Africa': Evidence for an early expansion of modern humans into Arabia. *Science* 331: 453–6.

Bell, A., Hinde, K., and Newson L. 2013. Who was helping? The scope for female cooperative breeding in early *Homo*. *PLoS ONE* 8: e83667.

Boyd, R. and Richerson, P.J. 2005. Solving the puzzle of human cooperation, in Levinson, S. (ed.), *Evolution and Culture*, MIT Press, Cambridge, MA, pp. 105–32.

Bramble, D.M. and Lieberman, D.E. 2004. Endurance running and the evolution of Homo. *Nature* 432: 345–52.

Calloway, E. 2016. Evidence mounts for interbreeding bonanza in ancient human species. *Nature*, http://www.nature.com/news/evidence-mounts-for-interbreeding-bonanza-in-ancient-human-species-1.19394.

Cann, R.L., Stoneking, M., and Wilson, A.C. 1987. Mitochondrial DNA and human evolution. *Nature* 325: 31–6.

Carmody, R.N. and Wrangham, R. 2009. The energetic significance of cooking. *Journal of Human Evolution* 57(4): 379–91.

Carto, S.L., Weaver, A.J., Hetherington, R. et al. 2009. Out of Africa and into an ice age: On the role of global climate change in the late Pleistocene expansion of early modern humans out of Africa. *Journal of Human Evolution* 56: 139–51.

Castañeda, I.S., Mulitza, S., Schefuß, E. et al. 2009. Wet phases in the Sahara/Sahel region and human expansion patterns in North Africa. *Proceedings of the National Academy of Sciences* 106: 20159–63.

Cieri, R., Churchill, S., Franciscus, R. et al. 2014. Craniofacial feminization, social tolerance, and the origins of behavioral modernity. *Current Anthropology* 55: 419–43.

Dunbar, R.I.M. 1998. The social brain hypothesis. *Evolutionary Anthropology* 6: 178–90.

Eleanor M.L. et al. 2018. Did our species evolve in subdivided populations across Africa, and why does it matter? *Trends in Ecology and Evolution*, DOI: https://doi.org/10.1016/j.tree.2018.05.005

Gagneux, P. and Varki, A. 2001. Genetic differences between humans and apes. *Molecular Phylogenetics and Evolution* 18(1): 2–13.

Gomez-Robles, A., Hopkins, W., Schapiro, S., and Sherwood, C. 2015. Relaxed genetic control of cortical organization in human brains compared with chimpanzees, *Proceedings of the National Academy of Sciences* 112(48): 14799–804.

Goodman, M., Porter, C.A., Czelusniak, J. et al. 1998. Toward a phylogenetic classification of primates based on DNA evidence complemented by fossil evidence. *Molecular Phylogenetics and Evolution* 9: 585–98.

Green R.E., Krause, J., Briggs, A.W. et al. 2010. A draft sequence of the Neandertal Genome. *Science* 328(5979): 710–22.

Gruss, L.T. and Schmitt, D. 2015. The evolution of the human pelvis: changing adaptations to bipedalism, obstetrics and thermoregulation. *Philosophical Transactios of the Royal Society B* 370: 1663.

Hare, B., Wobber, V., and Wrangham, R. 2012. The self-domestication hypothesis: Evolution of bonobo psychology is due to selection against aggression. *Animal Behaviour* 83: 573–85.

Harvey, P.H., Martin, R.D., and Clutton-Brock, T.H. 1987. Life histories in comparative perspective, in Smuts, B.B., Cheney, D.L., Seyfarth, R.M. et al. (eds) *Primate Societies*, University of Chicago Press, Chicago, IL, pp. 181–96.

Isler, K. and van Schaik, C.P. 2009. The expensive brain: A framework for explaining evolutionary changes in brain size. *Journal of Human Evolution* 57: 392–400.

Isler, K. and van Schaik, C.P. 2014. How humans evolved large brains: Comparative evidence. *Evolutionary Anthropology* 23: 65–75.

Kim, B.Y. and Lohmueller, K.E. 2015. Selection and reduced population size cannot explain higher amounts of Neandertal ancestry in East Asian than in European human populations. *American Journal of Human Genetics* 96(3): 454–61.

Lao, O., de Gruijter, J.M., van Duijn, K. et al. 2007. Signatures of positive selection in genes associated with human skin pigmentation as revealed from analyses of single nucleotide polymorphisms. *Annals of Humuman Genetics* 71: 354–69.

Liu, W., Martinon-Torres, M., Cai, Yan-jun et al. 2015. The earliest unequivocally modern humans in southern China. *Nature* 526: 696–9.

Manica, A., Amos, B., Balloux, F., and Hanihara, T. 2007. The effect of ancient population bottlenecks on human variation. *Nature* 448(7151): 246–8.

Maslin, M., Shultz, S., and Trauth, M. 2015. A synthesis of the theories and concepts of early human evolution. *Philosophical Transactions of the Royal Society B* 370.

McGrew, W. 2010. Chimpanzee technology. *Nature* 328: 579–80.

McPherron, S.P., Alemseged, Z., Marean, C.W. et al. 2010. Evidence for stone-tool-assisted consumption of animal tissues before 3.39 million years ago at Dikika, Ethiopia. *Nature* 466: 857–60.

Meyer, M. et al. 2012. A high coverage genome sequence from an archaic Denisovan individual. *Science* 338(6014): 222–6.

Meyer, M., Arsuaga, J.L., de Filippa, C. et al. 2015. Nuclear DNA sequences from the Middle Pleistocene Sima de los Huesos hominins. *Nature* 531: 504–7.

Noonan, J.P., Coop, G., Kudaravalli, S. et al. 2006. Sequencing and analysis of Neanderthal genomic DNA. *Science* 314(5802): 1113–18.

O'Connell, J.F., Hawkes, K., and Blurton Jones, N.G. 1999. Grandmothering and the evolution of *Homo erectus*. *Journal of Human Evolution* 36: 461–85.

Opie, C., Atkinson, Q.D., Dunbar, R.I.M. et al. 2013. Male infanticide leads to social monogamy in primates. *Proceedings of the National Academy of Sciences* 110: 13328–32.

Preuss, T.M. 2012. Human brain evolution: From gene discovery to phenotype discovery. *Proceedings of the National Academy of Sciences* 109: s10709–16.

Pruefer K., Racimo, F., Patterson, N. et al. 2014. The complete genome sequence of a Neanderthal from the Altai mountains. *Nature* 505(7481): 43–9.

Racimo, F., Sankararaman, S., Nielsen, R., and Huerta-Sanchez, E. 2015. Evidence for archaic adaptive introgression in humans. *Nature Reviews Genetics* 16(6): 359–71.

Reich, D., Green, R.E., Kircher, M. et al. 2010. Genetic history of an archaic hominin group from Denisova Cave in Siberia. *Nature* 468(7327): 1053–60.

Rosenberg, K. and Trevathan, W. 2002. Birth, obstetrics and human evolution. *BJOG* 109: 1199–206.

Sankararaman, S., Patterson, N., Li, H. et al. 2012. The date of interbreeding between Neandertals and modern humans. *PLoS Genetics* 8(10): e1002947.

Sankararaman, S., Mallick, S., Dannemann, M. et al. 2014. The genomic landscape of Neanderthal ancestry in present-day humans. *Nature* 507(7492): 354–7.

Tague, R.G. and Lovejoy, O. 1986. The obstetric pelvis of A.L. 288–1 (Lucy). *Journal of Human Evolution* 15: 237–55.

Tishkoff, S.A. and Williams, S.M. 2002. Genetic analysis of African populations: Dissecting human evolutionary history and complex disease. *Nature Reviews Genetics* 3(8): 611–21.

Vernot, B. and Akey, J.M. 2014. Resurrecting surviving Neandertal lineages from modern human genomes. *Science* 343(6174): 1017–21.

Vernot, B. and Akey, J.M. 2015. Complex history of admixture between modern humans and Neandertals. *American Journal of Human Genetics* 96(3): 448–53.

Wall, J.D., Yang, M.A., Jay, F. et al. 2013. Higher levels of Neanderthal ancestry in East Asians than in Europeans. *Genetics* 194(1): 199–209.

Wildman, D., Uddin, M., Liu, G. et al. 2003. Implications of natural selection in shaping 99.4% nonsynonymous DNA identity between humans and chimpanzees: Enlarging genus Homo. *Proceedings of the National Academy of Sciences* 100 (12): 7181–8.

CHAPTER 9: THE FUTURE OF HUMANITY

Books

Brownlee, D. and Ward, P. 2007. *The Life And Death Of Planet Earth: How Science Can Predict the Ultimate Fate of Our World*, Piatkus, London, p. 256.

Harari, Y.N. 2014. *Sapiens: A Brief History of Humankind*, Harvill Secker, London, p. 443.

Hoffman, P.T. 2015. *Why Did Europe Conquer the World?* Princeton University Press, Princeton, NJ, p. 272.

Leakey, R. and Lewin, R. 1996. *The Sixth Extinction: Biodiversity and its Survival*, Weidenfeld and Nicolson, London, p. 271.

Lewis, S.L. and Maslin, M.A., 2018. The Human Planet; How we caused the Anthropocene, Penguin, p. 465.

Maslin, M. 2014. *Climate Change: A Very Short Introduction*, 3rd edition, Oxford University Press, Oxford, p. 187.

Ruddiman, W.F. 2010. *Plows, Plagues, and Petroleum: How Humans Took Control of Climate*, Princeton University Press, Princeton, NJ, p. 240.

Journals and chapters

Barnosky, A.D., Matzke, N., Tomiya, S. et al. 2011. Has the Earth's sixth mass extinction already arrived? *Nature* 471: 51–7.

Chakrabarty, D. 2009. The climate of history: Four theses. *Critical Inquiry* 35(Winter): 197–222.

Chakrabarty, D. 2015. Anthropocene, in Wenzel, J. and Szeman, I. (eds), *Fuelling Culture: Energy, History, Politics*, Fordham University Press, New York.

Chakrabarty, D. 2015. The Anthropocene and the convergence of histories, in Hamilton, C., Bonneuil, C., and Gemenne, F. (eds), *The Anthropocene and the Global Environmental Crisis:*

Rethinking Modernity in a New Epoch, Routledge, Oxford, pp. 44–56.

Crutzen, P.J. and Stoermer, E.F. 2000. The Anthropocene. *IGBP Newsletter* 41(17): 17–18.

Dalby, S. 2007. Anthropocene geopolitics: Globalisation, empire, environment and critique. *Geography Compass* 1(1): 103–18.

Foley, S.F., Gronenborn, D., Andreae, M.O. et al. 2013. The Palaeoanthropocene: The beginnings of anthropogenic environmental change. *Anthropocene* 3: 83.

Hamilton, C. and Grinevald, J. 2015. Was the Anthropocene anticipated? *The Anthropocene Review* 2(1): 59–72.

Johnson, E. and Morehouse, H. 2014. After the Anthropocene: Politics and geographic inquiry for a new epoch. *Progress in Human Geography* 38(3): 439–56

Latour, B. 2015. Telling friends from foes in the time of the Anthropocene, in Hamilton, C., Bonneuil, C., and Gemenne, F. (eds), *The Anthropocene and the Global Environmental Crisis: Rethinking Modernity in a New Epoch*, Routledge, Oxford, 145–55.

Lewis, S.L. and Maslin, M.A. 2015. A transparent framework for defining the Anthropocene Epoch. *Anthropocene Review* <http://anr.sagepub.com/content/early/2015/05/15/2053019615588792>.

Lewis, S.L. and Maslin, M.A. 2015. Defining the Anthropocene. *Nature* 519: 171–80.

Maslin, M.A. and Lewis, S.L. 2015. Anthropocene: Earth system, geological, philosophical and political paradigm shifts. *The Anthropocene Review* 2(2): 108–16.

Palumbi, S.R. 2001. Humans as the world's greatest evolutionary force. *Science* 293: 1786–90.

Ruddiman, W.F., Ellis, E.C. et al. 2015. Defining the epoch we live in. *Science* 348: 38–9.

Steffen, W., Broadgate, W., Deutsch, L. et al. 2015. The trajectory of the Anthropocene: The Great Acceleration. *The Anthropocene Review* 2(1): 81–98.

Zalasiewicz, J., Waters, C.N., Williams, M. et al. 2015. When did the Anthropocene begin? A mid-twentieth century boundary level is stratigraphically optimal. *Quaternary International* 383: 196–203.

Zeder, M.A. 2011. The origins of agriculture in the Near East. *Current Anthropology* 52: s221–35.

INDEX

ORIGINS

The Scientific Story of Creation

Jim Baggott

978-0-19-870764-6 | Hardback | £25.00

'The collective mind of humanity has made extraordinary progress in its quest to understand how the current richness of the physical world has emerged, and Baggott with his characteristic lucidity and erudition, has provided an enthralling account of this wonderful and still unfolding intellectual journey.'

– Peter Atkins

'There are many different versions of our creation story. This book tells the version according to modern science', writes Jim Baggott. In *Origins*, he presents a unique version of the story in chronological sequence, from the Big Bang to the emergence of human consciousness 13.8 billion years later.

Cosmology, particle physics, chemistry, planetary geology, biology it is all here, explained with clarity, in one overarching narrative. And throughout, Baggott emphasizes that the scientific story is a work-in-progress, highlighting the many puzzles and uncertainties that still remain. We have a seemingly innate desire to comprehend our own place in the Universe. Jim Baggott helps us fulfil this desire, which is driven in part by simple curiosity but also by a deeper emotional need to connect ourselves meaningfully with the world which we call home.

THE IMPROBABLE PRIMATE

How Water Shaped Human Evolution

Clive Finlayson

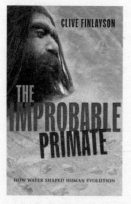

'Finlayson writes in a dry, clear, scholarly style which somehow accentuates the sheer improbability of humanity's long journey'
  – Brandon Robshaw, *Independent on Sunday*

Taking an ecological approach to our evolution, Clive Finlayson considers the origins of modern humans within the context of a drying climate and changing landscapes. Finlayson argues that environmental change, particularly availability of water, played a critical role in shaping the direction of human evolution, contributing to our spread and success.

978-0-19-874389-7 | Paperback | £10.99

CLIMATE CHANGE

A Very Short Introduction

Mark Maslin

978-0-19-871904-5 | Paperback | £7.99

Global warming is arguably the most critical and controversial issue facing the world in the twenty-first century. This *Very Short Introduction* provides a concise and accessible explanation of the key topics in the debate: looking at the predicted impact of climate change, exploring the political controversies of recent years, and explaining the proposed solutions. Fully updated for 2008, Mark Maslin's compelling account brings the reader right up to date, describing recent developments from US policy to the UK Climate Change Bill, and where we now stand with the Kyoto Protocol. He also includes a chapter on local solutions, reflecting the now widely held view that, to mitigate any impending disaster, governments as well as individuals must act together.

DIVIDED NATIONS

Why global governance is failing, and what we can do about it

---

Ian Goldin

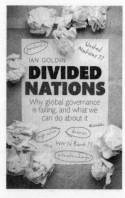

978-0-19-968903-3 | Paperback | £9.99

'Divided Nations is an absolutely remarkable book, which provides fresh and particularly useful theoretical as well as necessarily practical insights given the present challenges facing humanity.'

– Jean-Claude Trichet, former president of the European Central Bank

With rapid globalization, the world is more deeply interconnected than ever before. While this has its advantages, it also brings with it systemic risks that are only just being identified and understood. Rapid urbanization, together with technological leaps, such as the Internet, mean that we are now physically and virtually closer than ever in humanity's history.

We face a number of international challenges—climate change, pandemics, cyber security, and migration—which spill over national boundaries. It is becoming increasingly apparent that the UN, the IMF, the World Bank—bodies created in a very different world, more than 60 years ago—are inadequate for the task of managing such risk in the 21st century.

Ian Goldin explores whether the answer is to reform the existing structures, or to consider a new and radical approach. By setting out the nature of the problems and the various approaches to global governance, Goldin highlights the challenges that we are to overcome and considers a road map for the future.

PROSPERITY

Better Business Makes the Greater Good

Colin Mayer

978-0-19-882400-8 | Hardback | £16.99

'Here is the case for reinventing the corporation so that it serves human well-being. Colin Mayer shows both why an exclusive focus on shareholder value is damaging, and how purposeful changes could support trustworthy corporations that combine social and business benefits.'

– Baroness Onora O'Neill of Bengarve,
Emeritus Professor of Philosophy,
University of Cambridge

What is business for? Day one of a business course will tell you: it is to maximise shareholder profit. This single idea pervades all our thinking and teaching about business around the world but it is fundamentally wrong, Colin Mayer argues. It has had disastrous and damaging consequences for our economies, environment, politics, and societies.

In this urgent call for reform, Prosperity challenges the fundamentals of business thinking. It sets out a comprehensive new agenda for establishing the corporation as a unique and powerful force for promoting economic and social wellbeing in its fullest sense—for customers and communities, today and in the future.

First Professor and former Dean of the Säid Business School in Oxford, Mayer is a leading figure in the global discussion about the purpose and role of the corporation. In Prosperity, he presents a radical and carefully considered prescription for corporations, their ownership, governance, finance, and regulation. Drawing together insights from business, law, economics, science, philosophy, and history, he shows how the corporation can realize its full potential to contribute to economic and social wellbeing of the many, not just the few.

Prosperity tells us not only how to create and run successful businesses but also how policy can get us there and fix our broken system.

RIVER OF LIFE, RIVER OF DEATH

The Ganges and India's Future

Victor Mallet

978-0-19-878617-7 | Hardback | £20.00

'An extraordinary and fascinating combination of history, geography, environment, politics, religion, and much more. Written with affection for and understanding of a country of special importance. . . . Not just the story of an often difficult past but also of hope for a possible healthy and attractive future.'
— Nicholas Stern, IG Patel Professor of Economics and Government at LSE

India is killing the Ganges, and the Ganges in turn is killing India. The waterway that has nourished more people than any on earth for three millennia is now so polluted with sewage and toxic waste that it has become a menace to human and animal health.

Victor Mallet traces the holy river from source to mouth, and from ancient times to the present day, to find that the battle to rescue what is arguably the world's most important river is far from lost. As one Hindu sage told the author in Rishikesh on the banks of the upper Ganges (known to Hindus as the goddess Ganga)—'If Ganga dies, India dies. If Ganga thrives, India thrives. The lives of 500 million people is no small thing.'

Drawing on four years of first-hand reporting and detailed historical and scientific research, Mallet delves into the religious, historical, and biological mysteries of the Ganges, and explains how Hindus can simultaneously revere and abuse their national river.

Can they succeed in saving the river from catastrophe—or is it too late?

THROUGH A GLASS BRIGHTLY

Using Science to See Our Species as We Really Are

David P. Barash

978-0-19-067371-0 | Hardback | £14.99

Human beings have long seen themselves as the center of the universe, the apple of God's eye, specially-created creatures who are somehow above and beyond the natural world. This viewpoint—a persistent paradigm of our own unique self-importance—is as dangerous as it is false. In Through a Glass Brightly, noted scientist David P. Barash explores the process by which science has, throughout time, cut humanity 'down to size,' and how humanity has responded. A good paradigm is a tough thing to lose, especially when its replacement leaves us feeling more vulnerable and less special. And yet, as science has progressed, we find ourselves—like it or not—bereft of many of our most cherished beliefs, confronting an array of paradigms lost.

Barash models his argument around a set of 'old' and 'new' paradigms that define humanity's place in the universe. This new set of paradigms range from provocative revelations as to whether human beings are well designed, whether the universe has somehow been established with our species in mind (the so-called anthropic principle), whether life itself is inherently fragile, and whether Homo sapiens might someday be genetically combined with other species (and what that would mean for our self-image). Rather than seeing ourselves through a glass darkly, science enables us to perceive our strengths and weaknesses brightly and accurately at last, so that paradigms lost becomes wisdom gained. The result is a bracing, remarkably hopeful view of who we really are.

THE AGE OF EM

Work, Love, and Life when Robots Rule the Earth

Robin Hanson

978-0-19-881782-6 | Paperback | £9.99

'Robin Hanson integrates ferocious future forces: robotics, artificial intelligence, overpopulation, economic stagnation—and comes up with a detailed, striking set of futures we can have, if we think harder.'

– Gregory Benford,
two times Nebula award winner

Robots may one day rule the world, but what is a robot-ruled Earth like?

Many think the first truly smart robots will be brain emulations or ems. Scan a human brain, then run a model with the same connections on a fast computer, and you have a robot brain, but recognizably human.

Train an em to do some job and copy it a million times: an army of workers is at your disposal. When they can be made cheaply, within perhaps a century, ems will displace humans in most jobs. In this new economic era, the world economy may double in size every few weeks.

While human lives don't change greatly in the em era, em lives are as different from ours as our lives are from those of our farmer and forager ancestors. Ems make us question common assumptions of moral progress, because they reject many of the values we hold dear.

Read about em mind speeds, body sizes, job training and career paths, energy use and cooling infrastructure, virtual reality, aging and retirement, death and immortality, security, wealth inequality, religion, teleportation, identity, cities, politics, law, war, status, friendship and love.

This book shows you just how strange your descendants may be, though ems are no stranger than we would appear to our ancestors. To most ems, it seems good to be an em.